Architecting AWS with Terraform

Design resilient and secure Cloud Infrastructures
with Terraform on Amazon Web Services

Erol Kavas

‹packt›

BIRMINGHAM—MUMBAI

Architecting AWS with Terraform

Group Product Manager: Preet Ahuja

Publishing Product Manager: Niranjan Naikwadi

Senior Editor: Divya Vijayan

Technical Editor: Irfa Ansari

Copy Editor: Safis Editing

Project Coordinator: Ashwin Kharwa

Proofreader: Safis Editing

Indexer: Rekha Nair

Production Designer: Prashant Ghare

Marketing Coordinator: Rohan Dobhal

First published: December 2023

Production reference: 1071223

Published by Packt Publishing Ltd.

Grosvenor House

11 St Paul's Square

Birmingham

B3 1RB

ISBN 978-1-80324-856-1

www.packtpub.com

To my incredible wife, Mihrimah, whose steadfast support and unwavering belief in me have been my guiding light through every challenge and triumph. Your selfless dedication and willingness to help, no matter the circumstances, have been the foundation upon which I have built my dreams. You are my rock, my partner, and my inspiration. This book is a testament to your love and support.

– Erol Kavas

Contributors

About the author

Erol Kavas is a renowned multi-cloud expert and cloud evangelist in Canada, with over 20 years of industry experience. As a **Microsoft Certified Trainer** (**MCT**) Regional Lead and an AWS Ambassador, Erol is a recognized authority in cloud computing. His career spans roles such as solutions architect, security architect, enterprise architect, and delivery lead. Erol's passion lies in helping enterprises become future-proof, with cutting-edge cloud infrastructure and security solutions. He holds over 100 certificates in cloud, security, and project management. Currently, Erol serves as a director in the cloud and data team at PwC Canada.

I want to thank the people who have been close to me and supported me, especially my wife, Mihrimah, my kids, Esad and Azra, and my colleagues.

About the reviewers

Hamza Koc has accumulated over five years of expertise in software development, cloud infrastructure, and DevOps. He stands out for his profound understanding of major cloud platforms, particularly Microsoft Azure, AWS, and GCP. As a cloud and DevOps engineer, Hamza has streamlined code delivery processes to Kubernetes clusters, architected scalable cloud solutions, and led automation efforts using tools such as Terraform. Beyond his hands-on roles, Hamza is an esteemed Microsoft Azure and Terraform trainer, guiding professionals in mastering cloud technologies and practices. His achievements are further underscored by a robust collection of certifications from industry leaders, such as Microsoft and AWS.

I wish to express my deep appreciation to my mentors and colleagues for their constant guidance. My utmost gratitude goes to my family for their unwavering belief in my capabilities, and to the visionary teams I've collaborated with, who have been a constant source of inspiration.

Dr. Ramazan Atalay is a global cloud security expert. With a B.Sc. in physics from METU and a Ph.D. in physics from Georgia State University, his education underpins his dynamic career in cybersecurity.

Currently at Loblaw Inc., he is a full-time employee and a Microsoft Certified Trainer for top courses on cloud security and cybersecurity. His passion for mentoring extends to his role as a colleague in Loblaw's cybersecurity, network, and technology risk department.

Dr. Atalay is a leading figure in the cybersecurity domain, with a dedicated commitment to education and knowledge sharing.

Table of Contents

3

Building Your First Terraform Project   35

4

Discovering Best Practices for Terraform IaC Projects   53

Part 2: Become an Expert in Terraform with AWS

5

Planning and Designing Infrastructure Projects in AWS   67

6

Making Decisions for Terraform Projects with AWS 89

7

Implementing Terraform in Projects 101

8

Deploying Serverless Projects with Terraform 113

9

Deploying Containers in AWS with Terraform   123

Part 3: How to Structure and Advance Terraform in Enterprises

10

Leveraging Terraform for the Enterprise   143

Preface

In the ever-evolving world of cloud computing, the ability to automate and manage infrastructure efficiently is paramount. As organizations continue to migrate their workloads to the cloud, the need for a robust and scalable approach to infrastructure management becomes increasingly evident. Enter Terraform, a powerful **Infrastructure-as-Code (IaC)** tool that enables developers and operations teams to automate the provisioning and management of cloud resources with ease. When combined with the vast capabilities of **Amazon Web Services (AWS)**, Terraform becomes an indispensable tool for building, deploying, and managing complex cloud infrastructure.

Architecting AWS with Terraform: Design resilient and secure Cloud Infrastructures with Terraform on Amazon Web Services is a comprehensive guide that aims to equip you with the knowledge and skills needed to harness the full potential of Terraform on AWS. Whether you are a start-up looking to build a scalable cloud infrastructure, an enterprise seeking to improve the reusability and governance in your Terraform projects, or an individual exploring the world of IaC, this book has something for everyone.

Written by a 12x AWS-certified cloud and infrastructure automation expert, DevOps trainer, and AWS ambassador, this book takes you on a journey through the intricacies of Terraform and AWS, providing real-life implementation tips and best practices along the way. You will begin by understanding the patterns and anti-patterns of IaC and Terraform, learning how to avoid common mistakes and pitfalls. As you progress, you will discover the importance of planning and designing infrastructure projects in AWS, and how to make informed decisions for your AWS Terraform projects.

The book delves into the practical implementation of Terraform in various projects, including deploying serverless applications and containers in AWS. You will learn how to leverage Terraform for enterprise-level projects, build Git workflows for IaC and Terraform projects, and automate the deployment of Terraform projects. Additionally, you will explore how to govern AWS resources with Terraform and build a secure infrastructure using AWS Terraform.

By the end of this book, you will have a comprehensive understanding of Terraform and IaC, and you will be equipped with the knowledge and skills needed to build, manage, and deploy complex infrastructure on AWS. Whether you are a cloud engineer, DevOps engineer, developer, or architect responsible for writing or designing IaC to deploy resources in AWS, this book will serve as a valuable resource on your journey to mastering Terraform on AWS.

Welcome to the world of automated, scalable, and secure cloud infrastructure with Terraform and AWS. Let's get started!

Who this book is for

Architecting AWS with Terraform: Design resilient and secure Cloud Infrastructures with Terraform on Amazon Web Services is designed for a wide range of professionals who are involved in the design, development, and deployment of cloud infrastructure on AWS. This book is particularly suited for the following roles:

- **Cloud engineers**: If you are responsible for designing, deploying, and managing cloud infrastructure on AWS, this book will provide you with the tools and techniques needed to automate and scale your infrastructure using Terraform.

- **DevOps engineers**: As a DevOps engineer, you are at the intersection of development and operations. This book will help you streamline your infrastructure deployment processes, improve reusability, and implement best practices for IaC using Terraform.

- **Developers**: If you are a developer looking to deploy applications on AWS, this book will guide you through the process of automating your infrastructure deployment using Terraform, allowing you to focus on building and deploying your applications.

- **Architects**: As an architect, you are responsible for designing scalable and secure cloud infrastructure. This book will provide you with insights into planning and designing infrastructure projects in AWS, making informed decisions for AWS Terraform projects, and building a secure infrastructure using AWS Terraform.

- **IT professionals**: If you are an IT professional interested in exploring the world of cloud computing and IaC, this book will serve as a comprehensive introduction to Terraform on AWS.

Readers are expected to have a basic understanding of AWS and should have deployed resources from the AWS Management Console before. Familiarity with cloud concepts and general programming principles will be beneficial but is not required.

Whether you are just starting your journey in cloud computing or are an experienced professional looking to enhance your skills in infrastructure automation, this book will provide you with valuable insights and practical knowledge to master Terraform on AWS.

What this book covers

Chapter 1, Understanding Patterns and Anti-Patterns of IaC and Terraform, introduces the concepts of IaC and Terraform, highlighting the best practices and common pitfalls to avoid.

Chapter 2, How Not to Use IaC and Terraform, teaches you about the common mistakes made when using IaC and Terraform and how to avoid them, ensuring a smoother and more efficient infrastructure management process.

Chapter 3, Building Your First Terraform Project, helps you get started with Terraform by creating your first project. This chapter will guide you through the initial setup and configuration of a simple Terraform project on AWS.

Chapter 4, Discovering Best Practices for Terraform IaC Projects, explores the best practices for managing Terraform projects, including code organization, modularization, and versioning, to ensure efficient and maintainable IaC.

Chapter 5, Planning and Designing Infrastructure Projects in AWS, helps you understand the importance of planning and designing your AWS infrastructure projects, including considerations for scalability, security, and cost optimization.

Chapter 6, Making Decisions for Terraform Projects with AWS, teaches you how to make informed decisions for your AWS Terraform projects, considering factors such as resource selection, configuration, and deployment strategies.

Chapter 7, Implementing Terraform in Projects, dives into the practical implementation of Terraform in various projects, including deploying serverless applications and containers in AWS.

Chapter 8, Deploying Serverless Projects with Terraform, explores how to use Terraform to deploy serverless projects on AWS, including the setup and configuration of AWS Lambda functions and API Gateway.

Chapter 9, Deploying Containers in AWS with Terraform, teaches you how to deploy containerized applications on AWS using Terraform, including the setup and configuration of Amazon ECS and EKS.

Chapter 10, Leveraging Terraform for the Enterprise, helps you to understand how to use Terraform for enterprise-level projects, including best practices for managing large-scale Terraform projects and improving reusability and governance.

Chapter 11, Building Git Workflows for IaC and Terraform Projects, is where you will discover how to integrate Git workflows into your IaC and Terraform projects, enabling version control, collaboration, and automated deployment.

Chapter 12, Automating the Deployment of Terraform Projects, teaches you how to automate the deployment of your Terraform projects using **continuous integration and continuous deployment (CI/CD)** pipelines.

Chapter 13, Governing AWS with Terraform, explores how to use Terraform to govern your AWS resources, including managing access control, monitoring, and compliance.

Chapter 14, Building a Secured Infrastructure with AWS Terraform, helps you to understand how to build a secure infrastructure on AWS using Terraform, including best practices for network security, data protection, and identity and access management.

Chapter 15, Perfecting AWS Infrastructure with Terraform, teaches you how to achieve perfect AWS infrastructure with Terraform, including optimizing performance, reliability, and cost effectiveness.

Conventions used

There are a number of text conventions used throughout this book.

`Code in text`: Indicates code words in text, database table names, folder names, filenames, file extensions, pathnames, dummy URLs, user input, and Twitter handles. Here is an example: "Mount the downloaded `WebStorm-10*.dmg` disk image file as another disk in your system."

A block of code is set as follows:

```
resource "azurerm_sql_database" "main" {
    name = "primary-instance"
    settings {
        tier = " "
}
```

When we wish to draw your attention to a particular part of a code block, the relevant lines or items are set in bold:

```
variable "readers" {
    description = "..."
    type = list
    default = []
}
```

Any command-line input or output is written as follows:

```
brew upgrade hashicorp/tap/terraform
```

Bold: Indicates a new term, an important word, or words that you see onscreen. For instance, words in menus or dialog boxes appear in **bold**. Here is an example: "Select **System info** from the **Administration** panel."

> **Tips or important notes**
> Appear like this.

Get in touch

Feedback from our readers is always welcome.

General feedback: If you have questions about any aspect of this book, email us at customercare@ packtpub.com and mention the book title in the subject of your message.

Errata: Although we have taken every care to ensure the accuracy of our content, mistakes do happen. If you have found a mistake in this book, we would be grateful if you would report this to us. Please visit www.packtpub.com/support/errata and fill in the form.

Piracy: If you come across any illegal copies of our works in any form on the internet, we would be grateful if you would provide us with the location address or website name. Please contact us at copyright@packtpub.com with a link to the material.

If you are interested in becoming an author: If there is a topic that you have expertise in and you are interested in either writing or contributing to a book, please visit authors.packtpub.com.

Share Your Thoughts

Once you've read *Architecting AWS with Terraform*, we'd love to hear your thoughts! Scan the QR code below to go straight to the Amazon review page for this book and share your feedback.

https://packt.link/r/1803248564

Your review is important to us and the tech community and will help us make sure we're delivering excellent quality content.

Download a free PDF copy of this book

Thanks for purchasing this book!

Do you like to read on the go but are unable to carry your print books everywhere?

Is your eBook purchase not compatible with the device of your choice?

Don't worry, now with every Packt book you get a DRM-free PDF version of that book at no cost.

Read anywhere, any place, on any device. Search, copy, and paste code from your favorite technical books directly into your application.

The perks don't stop there, you can get exclusive access to discounts, newsletters, and great free content in your inbox daily

Follow these simple steps to get the benefits:

1. Scan the QR code or visit the link below

https://packt.link/free-ebook/9781803248561

2. Submit your proof of purchase

3. That's it! We'll send your free PDF and other benefits to your email directly

Part 1:
Introduction to IAC and
Terraform in AWS

In this initial section, we establish the essential groundwork for mastering Terraform on AWS. We explore the fundamental concepts of **Infrastructure as Code (IaC)** and Terraform, including both effective patterns and common pitfalls. You'll learn how to avoid typical mistakes and build your first Terraform project on AWS with confidence. We'll also delve into best practices for managing Terraform projects, ensuring you have a solid foundation for efficient and maintainable IaC. By the end of this part, you'll be well-equipped to start deploying cloud infrastructure on AWS using Terraform.

This part contains the following chapters:

- *Chapter 1, Understanding Patterns and Anti-Patterns of IaC and Terraform*
- *Chapter 2, How Not to Use IaC and Terraform*
- *Chapter 3, Building Your First Terraform Project*
- *Chapter 4, Discovering Best Practices for Terraform IaC Projects*

1

Understanding Patterns and Antipatterns of IaC and Terraform

In an ever-evolving digital landscape, the seamless integration of development and operations has become a necessity for organizations seeking to achieve unparalleled efficiency and agility. The opening chapter of this book delves into the fascinating world of **Infrastructure as Code (IaC)** and **Terraform**, unraveling the key principles, patterns, and anti-patterns that underpin this transformative approach. With a keen focus on idempotency, immutability, and an array of best practices, this chapter illuminates the path to robust, secure, and compliant infrastructure management. As we embark on this captivating journey, we'll explore the intricacies of IaC projects, examine the challenges they present, and unearth invaluable strategies to conquer them. By the end of this chapter, you'll possess a solid foundation to make informed decisions about the life cycle of your infrastructure and harness the true potential of IaC and Terraform.

We'll cover these main topics in this chapter:

- Introducing IAC
- Patterns and practices of IaC
- How to handle IaC projects
- How to make decisions about IaC projects

Introducing IaC

IaC refers to the process of managing and provisioning computing infrastructure through machine-readable definition files instead of relying on interactive configuration tools or physical hardware setups.

IaC leverages coding techniques that have been tried and tested in software systems, extending their application to infrastructure. It is one of the key DevOps practices that enable teams to deliver infrastructure and software rapidly and reliably at scale. Having a fast and dependable infrastructure provisioning mechanism is essential for organizations that want to achieve continuous delivery for their applications.

In IaC, a declarative language is typically used to describe the desired state of a system, as well as the steps required to bring it into compliance with that state. The IaC tool then uses these descriptions to construct and manage the necessary steps automatically, transitioning the system from one state to another. As a result, IaC enables organizations to automate processes such as resource installation, configuration, deployment, scaling, updating, and deletion in their IT infrastructures.

Key principles of IaC

There are two key principles of IaC, which we will gain an understanding of in this section.

Idempotency

Idempotency is a characteristic of certain operations in mathematics, programming languages, and computer science. It refers to the property where applying these operations multiple times produces the same result without altering it except for generating identical copies.

In the context of IaC, idempotency means that regardless of the starting state and the number of times the IaC is executed, the end state remains the same. This simplifies the infrastructure provisioning process and minimizes the likelihood of inconsistent outcomes. This property offers several advantages for operations, such as the capability to roll back changes and retry them in case of failure.

One way to achieve idempotency is by using a stateful tool such as Terraform. With Terraform, you can specify the desired end state of the infrastructure, and the tool will handle the process of reaching that state.

Immutability

Configuration change management is an important topic for infrastructure provisioning. For success, we need a powerful change management recording system that records all changes made to the infrastructure, and it includes details about why those changes were made, who was responsible for them, when they were implemented, and so on.

Configuration drift can pose a significant challenge to infrastructure management. It arises when changes are made to the infrastructure without proper documentation, causing different environments to diverge in ways that are difficult to replicate. This problem is particularly prevalent in mutable infrastructures that are active for extended periods.

The consequence of configuration drift can be severe, leading to inconsistent performance and stability and security issues in the infrastructure. Since it is difficult to reproduce the exact conditions that led to the drift, troubleshooting such problems can be time-consuming and error-prone.

Immutable infrastructure is a technique for constructing and managing infrastructure in a dependable, repeatable, and foreseeable manner. This approach offers several advantages over traditional IT environment management methods. Rather than altering the existing infrastructure, immutable infrastructure involves replacing it with a new one. By provisioning fresh infrastructure each time, the approach ensures that the infrastructure remains reproducible and free from configuration drift over time.

Immutable infrastructure also provides scalability when provisioning infrastructure in cloud environments.

Now that we know what IaC is and what its key principles are, let's look at the patterns of IaC.

Patterns and practices of IaC

Diving into the world of IaC, it is essential to uncover the patterns and practices that form the backbone of efficient and reliable implementations. In this section, we will explore the fundamental building blocks that contribute to the success of IaC, ensuring a comprehensive understanding of its best practices and a solid foundation for your IaC journey.

Source control and VCS

It is crucial to keep all aspects of your infrastructure, including the smallest scripts and pipeline configurations, in source control or **version control systems** (**VCSs**). A version control system is a tool that manages and tracks changes to documents, programs, and other collections of information, often used in software development to maintain a history of code changes.

This practice ensures that you have a record of all changes made to your infrastructure, regardless of how minor they may be. It also simplifies the process of tracking ownership and the history of changes to your infrastructure configurations.

Furthermore, it is important to make the infrastructure code accessible to all members of your organization, including those who do not directly work on the IaC code base. This visibility provides a better understanding of how the infrastructure is provisioned and enables quick troubleshooting of any issues that arise. By reviewing the code, users can gain a deeper understanding of how the infrastructure operates, and even contribute to the development of the infrastructure if they choose to do so.

The visibility and understanding of the applications running on your infrastructure are crucial for managing a successful IT infrastructure. By having a good grasp of how the applications function, you can optimize their performance and ensure that they operate efficiently. By keeping the infrastructure code accessible to all, you can ensure that your entire organization can contribute to maintaining and improving the infrastructure, ultimately leading to better outcomes for your business.

Modules and versions

Creating reusable modules in IaC tools helps with maintenance, readability, and ownership. It keeps changes small and independently deployable and reduces the effect radius.

Refactoring IaC is difficult compared to application development, particularly for critical pieces such as DNS records, network configurations, databases, and so on.

In many organizations, team structures and responsibilities are different, so it will make more sense to separate multiple layers of infrastructure and assign governance to the respective teams. In some cases, there might be some more separated layers needed for cross-functional teams managing both infrastructure and application development.

The following diagram illustrates an example of Amazon EKS deployments, featuring multiple modules for each infrastructure layer and their respective governors. It is important to note that the modules and layers depicted in this diagram may differ depending on your specific setup.

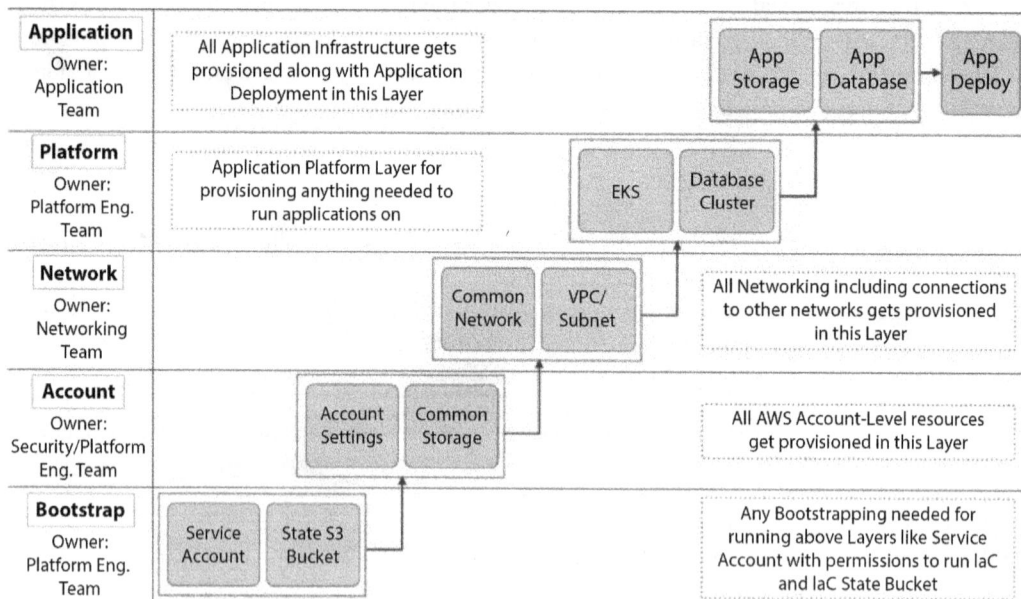

Figure 1.1 – EKS deployment workflow

Versioning for modules is quite important to provide support for multiple versions of services that can operate without breaking the existing production resources.

Documentation

IaC minimizes the need for extensive documentation for infrastructure since everything is codified and stated as a declarative manifest. However, some documentation is needed for better infrastructure provisioning so that consumers can understand and improve the current modules and templates.

Documentation can be challenging to manage, much like code. It is critical to provide sufficient documentation to convey the intended message effectively. However, having more documentation does not necessarily equate to better-quality documentation. In fact, outdated documentation can be more detrimental than having no documentation at all.

IaC documentation must live close to the code. Keep it close so that everyone can update the documentation without unnecessary effort and difficult steps. If you can build good governance automation, documentation creation or updates can be easily tracked and enforced.

An effective approach to managing documentation for IaC is to include a README file within the same repository as the code, rather than using an external platform such as Confluence or a wiki. This approach facilitates updating the documentation during the same commit as the code changes, which is particularly useful as a reminder during the pull request process.

It is also ideal to leverage automated tools to generate documentation from the code or use tests as documentation. By doing so, you can ensure that the documentation stays in sync with the code, reducing the likelihood of inconsistencies and outdated information. This approach can also streamline the documentation process, reducing the need for manual documentation efforts and enabling faster iterations.

Testing

Software testing is the process of executing a program or application with the intent of finding errors. Testing can be done at various levels, from unit testing to integration testing to system testing and acceptance testing.

IaC development is not an easy task. There are many different aspects and considerations that need to be taken into account before, during, and after the development process. One of these considerations is how to test your IaC. Let's provide you with a basic understanding of the various levels of testing that you need to think about when developing your IaC:

- **Static code and analysis**

  Running quick tests as frequently as possible is crucial for obtaining prompt feedback during the development process. This approach is especially effective when performed on your local machine. There are various integrations available that can automate this process and trigger tests automatically when you save a file in your text editor or IDE.

To perform static analysis, you can use specialized tools such as Terraform Validate or TFLint. These tools enable you to identify issues in your code and configurations promptly, reducing the likelihood of errors and inconsistencies in your infrastructure. By incorporating quick testing and static analysis into your development process, you can streamline the testing process and improve the reliability of your infrastructure.

- **Unit testing**

   Since many IaC tools, such as Terraform and Ansible, operate on a declarative model, unit testing may not always be necessary. However, in some cases, unit tests can be beneficial, particularly when conditionals or loops are involved.

   While unit testing may not always be required for IaC, incorporating it where necessary can help to catch potential issues early on in the development process, improving the overall quality of your infrastructure.

- **Integration testing**

   One essential step in ensuring the reliability of your infrastructure is to perform validation testing. This involves provisioning resources in a test environment and verifying whether specific requirements are met. It is crucial to avoid writing tests for things that are already covered by your IaC tool, particularly when working with declarative code.

   For example, instead of verifying whether the policies specified in IaC were applied, you should write automated tests to ensure that none of your S3 buckets are public. Similarly, you can test that only specific ports are open across all of your EC2 instances. To perform these tests, you can provision an ephemeral environment that you can later tear down.

   Depending on the duration of these tests, you may want to run them after every commit or as nightly builds. By incorporating validation testing into your development process, you can catch potential issues early on, reduce the risk of errors, and ensure the overall reliability of your infrastructure.

- **Smoke tests**

   An additional approach to testing is to provision an environment, deploy a dummy application, and run quick smoke tests to verify that the application has been deployed correctly. Using a dummy application can be helpful in testing scenarios that your actual application may encounter but are not configured for production.

   For example, if your application connects to an externally hosted database, you should attempt to connect to it in your dummy application. By doing so, you can gain confidence that the infrastructure you are provisioning is capable of supporting the applications you intend to run on it.

As these tests can be time-consuming, it is advisable to run them after provisioning a new environment and periodically thereafter. By leveraging this testing approach, you can ensure that your infrastructure is capable of supporting your application's requirements and minimize the risk of errors or issues arising during deployment.

Security and compliance

The definition of IaC is to provide an abstraction layer between the physical infrastructure and the applications that run on top of it. This is done by separating the hardware from the software and by abstracting out all of the tasks that are required to manage the hardware.

IaC can be used by companies for compliance purposes, such as HIPAA, SOX, PCI DSS, and so on. It can also be used for security purposes, such as preventing unauthorized access to data or preventing hackers from accessing sensitive information.

Let's look at important details of security and compliance.

Identity and access management

Implementing a strong **Identity and Access Management (IAM)** strategy is essential for safeguarding both your IaC and the infrastructure it provisions. One effective approach is to use **Role-Based Access Control (RBAC)** for IaC, which can significantly reduce the overall attack surface.

By leveraging RBAC, you can grant just enough permission to your IaC to perform the necessary operations while preventing unauthorized access. This approach helps to minimize the risk of errors or malicious activity, improving the overall security of your infrastructure.

Secrets management

When working with IaC, it is common to require secrets to provision infrastructure. For example, if you are provisioning resources in AWS, you will need valid AWS credentials to connect to it. It is crucial to ensure that you use a reliable secret management tool, such as HashiCorp Vault or AWS Secrets Manager, to manage these sensitive credentials.

In cases where you need to store or output secrets in the state file (although it is advisable to avoid doing so), it is essential to encrypt them to prevent unauthorized access. By encrypting secrets stored in the state file, you can mitigate the risk of exposure in the event of a security breach or unauthorized access.

Security scanning

Performing security scans after provisioning or making changes to infrastructure in a lower or ephemeral environment can help mitigate potential security issues in production. Leveraging tools such as CIS Benchmarks and Amazon Inspector can be effective in identifying common vulnerabilities or exposures and ensuring adherence to security best practices.

By conducting security scans, you can catch potential security issues early on in the development process and prevent them from being carried over to production. This approach helps to minimize the risk of security breaches and protect sensitive data and infrastructure.

Compliance

Compliance requirements are a critical consideration for many organizations, particularly in highly regulated industries such as healthcare or finance. These industries are subject to stricter requirements, including HIPAA, PCI, GDPR, and SOX, to name a few. Traditionally, compliance teams conducted manual checks and filled in paperwork to ensure adherence to these requirements.

However, automation tools such as Chef InSpec or HashiCorp Sentinel can help streamline compliance requirements and improve efficiency. By automating compliance checks, you can run them more frequently and identify issues much faster. For instance, you can incorporate compliance tests into your IaC pipeline by provisioning an ephemeral environment and running tests every time you modify your IaC code. This approach enables you to catch potential compliance issues early on and rectify them before they impact production systems.

How to handle IaC projects

In today's fast-paced digital landscape, IaC has become a critical consideration for organizations of all sizes. With IaC, developers can create the machines or resources required to run their applications easily, saving time and effort in the process. As your organization scales, IaC can help your developers focus on solving more complex problems, rather than getting bogged down in manual resource configuration.

However, it can be challenging to ensure identical, error-free, secure, and compliant configurations across different environments. This is where IaC comes in. By defining your infrastructure as code, you can make changes or add new resources by updating a piece of code, and the IaC tool will handle the configuration for you.

By adopting IaC, organizations can improve agility, speed, and consistency in resource provisioning and configuration. This enables developers to focus on delivering high-quality applications, while operations teams can manage infrastructure at scale with greater ease and efficiency.

Let's have a look at the challenges we can face.

IaC principles

At the heart of IaC is the concept of defining your infrastructure in code. By using a declarative syntax, you define the desired final state of your infrastructure, and the IaC tool takes care of the underlying dependency resolution and resource launching steps.

To keep track of changes made to your infrastructure, you can store this code in a VCS. This not only provides you with an audit trail of who made changes but also enables you to revert to a previous version if needed.

Automated quality, compliance, and security tests can also be run on your infrastructure, allowing you to verify its compliance without investing days or weeks of effort.

By adopting IaC, your developers can avoid the tedious and error-prone task of manually defining steps or scripts to launch and configure resources. Tools such as Terraform and CloudFormation are widely used to achieve these tasks, enabling organizations to achieve greater agility, scalability, and consistency in infrastructure management.

Version control systems for IaC

It is important to store your IaC in a VCS alongside your application code. This allows for easy collaboration among developers and a clear understanding of the entire code base.

VCSs also offer a simple way to track and audit changes made to the code base, including infrastructure changes. By using pipeline features within a VCS, such as those available in GitHub or GitLab, you can enforce policies and ensure that changes meet the necessary criteria before they are deployed to production.

Some common use cases of IaC

IaC is commonly used to launch infrastructure across various cloud providers, as well as for provisioning machines upon launch. Popular tools for provisioning with IaC include Chef, Ansible, and Puppet, while Terraform and CloudFormation are commonly used for infrastructure provisioning.

IaC can also be used to deploy applications, such as with Kubernetes, by leveraging tools such as Jenkins or Ansible. In upcoming chapters, we will delve further into using IaC with Kubernetes.

Challenges and best practices with IaC

IaC provides great benefits in terms of operability and maintainability, but it also brings challenges that need to be addressed to ensure the security and stability of your infrastructure.

Adoption within the team

Integrating IaC into your organization can present a learning curve and a change in processes. Your team may need to become familiar with the language used to write IaC code and develop pipelines to execute the code. If your team is accustomed to making changes from cloud consoles and is operation-centric, transitioning to IaC can be a significant shift for them.

You can see huge, powerful resistance to learning new technologies or practices. Be ready to fight, and always be an evangelist of infrastructure automation, security, and compliance.

Configuration drift

At the start of an IaC journey, developers may not always know what changes are required for infrastructure provisioning and may opt to make changes manually via the console. This can lead to configuration drift, where the deployed infrastructure does not match the code definition, potentially causing outages or issues with future updates. To prevent this, it is important to educate the team on the consequences of manual changes and discourage their use.

To further mitigate the risk of configuration drift, you can build automation to detect drifts and ensure that only authorized personnel have access to make changes in critical environments. This can help ensure that your infrastructure remains consistent and secure.

Security

When using open source modules in your IaC pipeline, it is important to ensure that they are secure and free of vulnerabilities. Before using any open source project, it is recommended to verify that it is safe to use.

To maintain a high level of security, it is essential to establish static code analysis pipelines and continuously scan open source modules. This way, any vulnerabilities can be detected and addressed promptly.

Human factors

To prevent misconfigurations from entering production, it is crucial to catch validation errors that may be introduced when a developer makes changes. With Terraform, you can easily implement a validation step using the Terraform plan functionality. It is essential to have a full understanding of the plan outputs before applying them to ensure that no unexpected changes are made to your infrastructure.

Side effects of automation

In IaC, a lot of code will be reused as you automate infrastructure creation. However, any small misconfiguration can propagate across a large set of resources very easily. Therefore, it's crucial to catch these errors during the pipeline verification stage.

To prevent unexpected changes to existing resources, always use versioning when updating modules.

Keeping up to date with cloud providers

Changes to cloud providers' APIs and policies can affect your existing infrastructure, which means that you need to update your tools and code. This can be especially difficult if you're using open source tools, as updates may not be immediately available. If there is a delay in releasing changes, it can result in incorrect permissions or issues with provisioning access to machines if the RBAC API changes. Therefore, it's essential to keep your tools and code up to date with the latest API changes and policies to ensure your infrastructure continues to function properly.

Maintainability and traceability

Having a well-defined procedure for promoting infrastructure changes to the production environment and assigning responsibilities is crucial to ensure that all changes are properly verified. This helps to avoid chaos and maintainability issues on the VCS side.

Furthermore, traceability is an added advantage of using VCSs as all changes are logged and can be easily tracked. For instance, Git provides the Git log command and commit history to view all changes made to the code.

RBAC

Many IaC tools, including Terraform, lack an intrinsic RBAC feature, a crucial element that governs who has permission to access, manage, and execute specific resources and operations. In the absence of native RBAC, these tools are dependent on the underlying platform or VCS where the code resides. Consequently, it's assumed that individuals executing the code possess the requisite permissions, transferring the onus of managing and enforcing RBAC to the VCS. This can involve setting up specific access controls, permissions, and restrictions within the VCS to ensure that sensitive and critical infrastructure configurations are only accessible and executable by authorized personnel, thereby maintaining security and compliance standards.

VCS and proper approval flows

It is essential to implement version control in your IaC workflow to maintain control of your code, track changes, and facilitate auditing. It is also important to establish a process where changes cannot be merged into production without proper approval and validation. One option is to incorporate validations into the **Continuous Integration (CI)** process of GitHub or GitLab. By treating your IaC code like any other application code, you can ensure that your infrastructure is an integral part of your overall system.

Handling secrets properly

You need to manage two types of secrets in your IaC pipeline. The first type of secret is used to create resources in the cloud, and only the admin of the repository should have access to them. For this purpose, you can use a secret variable in GitHub or GitLab.

The second type of secret is generated when the code is executed, such as the password for an IAM user in AWS. It's crucial to ensure that these secrets are not getting logged anywhere and are securely transmitted to users.

Immutable infrastructure

Consider applying the principle of immutable infrastructure if you need to make changes to your infrastructure. This approach involves creating a new machine with the required changes and replacing the old machine with the new one, instead of modifying the existing machine. By doing so, you can ensure that your changes are in line with the code, and there are no snowflake server states. The concept behind immutable infrastructure is to manage machines entirely through code, and no manual changes should be made.

Validations and checks

By implementing checks and validations in the CI pipeline, you can catch security issues and misconfigurations on the left side of the pipeline. This helps increase the frequency of the development cycle and maintain the security of each release.

Infrastructure as code and Kubernetes

Using the same principles as IaC, you can deploy your application on Kubernetes. Kubernetes objects are declarative files that can be defined and stored in a code repository. These files can then be applied to a Kubernetes cluster using a controller to deploy your application.

Conclusion

Despite the many advantages of IaC, there are also several challenges that must be addressed to ensure the success of the implementation. These include the need for proper validations and checks, as well as a well-established process to avoid security lapses that can lead to increased costs and compromised environments.

Fortunately, the emerging practice of GitOps combined with IaC enables faster and safer rollout of changes, resulting in quicker deployment cycles and large-scale auditing. IaC is not only the present but also the future of managing infrastructure, applications, and tooling, and its adoption is highly recommended for reducing operational costs.

By using IaC tools, organizations can achieve the same level of productivity and efficiency with fewer personnel, making it an attractive option for businesses looking to optimize their resources.

How to make decisions about IaC projects

IaC is a set of best practices for developers to document and configure their software infrastructure in a repeatable way.

IaC is not just about configuration management and deployment; it also provides the ability to manage infrastructure with code. The code can be used to automate activities such as application deployment, configuration management, and continuous delivery.

Here are a few plus points to consider:

- It is easy for developers to get started with IaC because the documentation is available in a single place
- It allows for more efficient collaboration between development teams by providing an easy way to share configurations with other members of the team
- It reduces errors in configuration management by making them easier to reproduce

Let's have a look at the decision points that will improve the maturity level of IaC projects.

The decision about where to store your code

Storing IaC files using a VCS is essential for tracking changes and collaboration. While any cloud storage system can be used, Git has become the de facto standard for IaC versioning. Originally designed for storing code, Git can be used as the primary source for deploying infrastructure code. Several solutions, such as GitHub, GitLab, and Bitbucket, offer free SaaS for public repositories, while community editions can be self-hosted. Using Git should be a basic skill set for any developer or cloud or DevOps engineer looking to start an IaC project successfully.

The decision about how to structure your code

Once you have chosen where to store your IaC code, the next step is deciding on how to structure it. The structure you choose will depend on the complexity of your organization and IT environment. There are several options, including using a mono-repo for all your IaC code, having a separate repository for each tool or language used, or having a repository for each application server or infrastructure type.

In addition, you need to determine a branching strategy that works well for your team. It's essential to discuss and agree on this with your team to ensure everyone is on the same page.

It's recommended to start with a simple structure and evolve it over time based on your needs. Alternatively, you can put more thought into the structure beforehand to prevent potential rework later. Whatever structure you choose, make sure it's easily adoptable by all team members. Create clear documentation on the structure and decision-making process so that new team members can quickly understand and start contributing effectively.

The decision about how to run your code

To gain better control over your infrastructure, it is recommended to use a CI/CD tool such as Jenkins, GitLab CI, or GitHub Actions to run your IaC. With these tools, you can trigger jobs manually, via webhooks or on a schedule, and have a record of every job that has run. Additionally, the jobs run from an agent can be pre-configured with the necessary tools, reducing the chances of errors due to different tool versions. It is important to choose the right tool that fits your needs and configure it properly to ensure its effectiveness.

The decision about how to handle your secrets

When provisioning automated infrastructure, it is crucial to store secrets such as database passwords and logins securely. It is not advisable to store them in your repositories, even if the repository is only accessible within your own network and protected with multi-factor authentication.

When using Git tools, all the credentials are copied to your machines and the machines of your team members when they clone the repository, making them vulnerable to security breaches.

A better solution is to use a vault system that can encrypt your secrets and inject them as environment variables during the runtime of your pipeline. It is ideal to have security enabled on multiple layers, so even if one layer is breached, there is a second line of defense to protect your sensitive information.

The decision about a common set of tools

To kickstart IaC projects effectively, it's important for the team to agree on a consistent set of tools. While there may be several ways to achieve the same objective, it's beneficial to explore simpler, quicker, or more cost-effective methods. Using a common toolset makes it easier to share and reuse building blocks. Striking a balance between granting engineers the freedom to experiment with new tools and standardizing on a common set of tools is crucial. Certain tools work well in tandem, while others don't, and paying for redundant licenses is generally not a good idea.

The decision about the level of pipelines

When using pipelines to run your IaC, there are various methods to achieve the same outcome. It's essential to use a naming convention and provide clear descriptions to help others understand the purpose of a pipeline. You can consider dividing a pipeline into multiple stages, so you have the flexibility to rerun or skip a stage depending on the type of deployment. Then, decide whether you want to enforce mandatory reviews, require approval from a manager, or give developers the liberty to deploy themselves during go-live.

The decision about the life cycle of your infrastructure

The level of testing and validation required for a proof-of-concept script versus code developed for large-scale deployment is significantly different. Robust code requires more comprehensive testing and validation efforts, which requires additional time and resources.

In an ever-evolving world, infrastructure must also be adaptable to changes such as security updates, service improvements, and new service types. While using SaaS/PaaS services can reduce the maintenance workload, it comes at a cost. Furthermore, even these services will evolve over time, necessitating engineering efforts to keep up. There are various strategies and practices available to simplify this process, each with its own benefits and drawbacks. It's important to determine the approach that works best for your specific situation.

Summary

This first chapter on understanding patterns of IaC and Terraform covered the key principles of IaC, such as idempotency and immutability. The chapter also discussed various patterns and practices of IaC, including source control, modules, versions, documentation, and testing. The chapter also covered security and compliance concerns, such as IAM, RBAC, secret management, security scanning, and compliance.

It also provided guidance on how to handle IaC projects and the decisions involved in starting IaC projects. Additionally, the chapter highlighted the challenges and best practices of IaC, including the importance of standardizing toolsets, naming conventions, and clear descriptions, and the need for a proper process for approvals and validation in the CI pipeline.

Overall, this chapter provided a comprehensive overview of the principles and best practices of IaC and highlighted the importance of adopting these practices to improve the agility, efficiency, and security of infrastructure management.

2

How Not to Use IaC and Terraform

Infrastructure as Code (IaC) tools such as Terraform have revolutionized the way we manage cloud infrastructure, offering powerful capabilities for automating the deployment and configuration of complex environments. However, despite the benefits, there are still many situations where IaC and Terraform can be misused or overused, leading to inefficiencies, errors, and security vulnerabilities.

In this chapter, we'll explore some basic Terraform commands, then compare Terraform and CloudFormation.

We'll cover these main topics in this chapter:

- Terraform architecture and workflow
- To compare with the others

Terraform architecture and workflow

To make the most of Terraform and use it effectively to manage your cloud infrastructure, it's important to have a good understanding of its architecture and workflow. Terraform follows a declarative approach to infrastructure management, where you describe the desired state of your infrastructure using configuration files, and it takes care of provisioning and configuring the necessary resources to achieve that state.

In this chapter, we'll explore the key components of Terraform's architecture and workflow, and provide a high-level overview of how Terraform works to manage infrastructure on cloud platforms such as AWS, Azure, and Google Cloud. Understanding these concepts will help you write more effective Terraform code and troubleshoot issues more easily.

Architecture

Terraform is a tool that allows developers to describe the infrastructure and run it in any environment. Terraform has many use cases, such as building a whole data center or just a single server or resource:

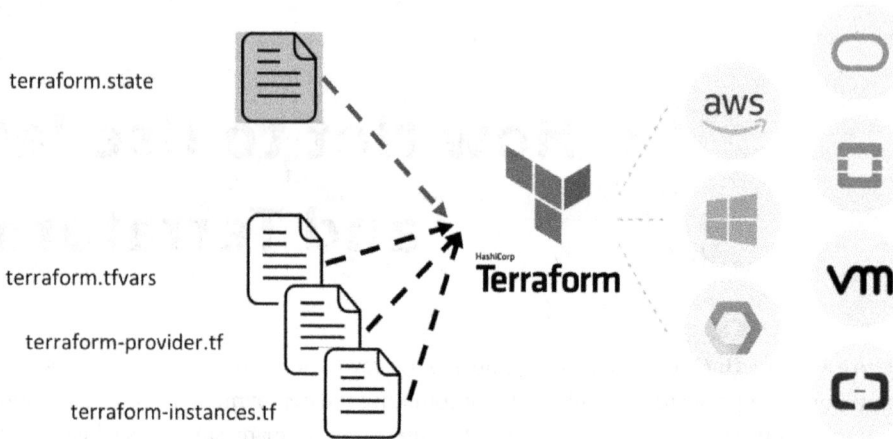

Figure 2.1 – Terraform – multi-cloud

Terraform is a powerful infrastructure management tool that enables safe and efficient building, modification, and versioning of infrastructure. It is capable of managing both local and remote infrastructure, making it an ideal choice for distributed teams collaborating on a project from different geographical locations. With Terraform, teams can easily work together on complex infrastructure projects, regardless of their physical location, while ensuring that the infrastructure remains consistent and up to date.

The Terraform architecture is quite simple; it is composed of just four components:

- Providers
- Modules
- Resources
- Templates

The modules are used to define the infrastructure and its components. The providers are responsible for how these modules should be executed in different environments. Resources are what we need to build our infrastructure, and templates are used to describe them easily. The following diagram outlines the technical flow for Terraform.

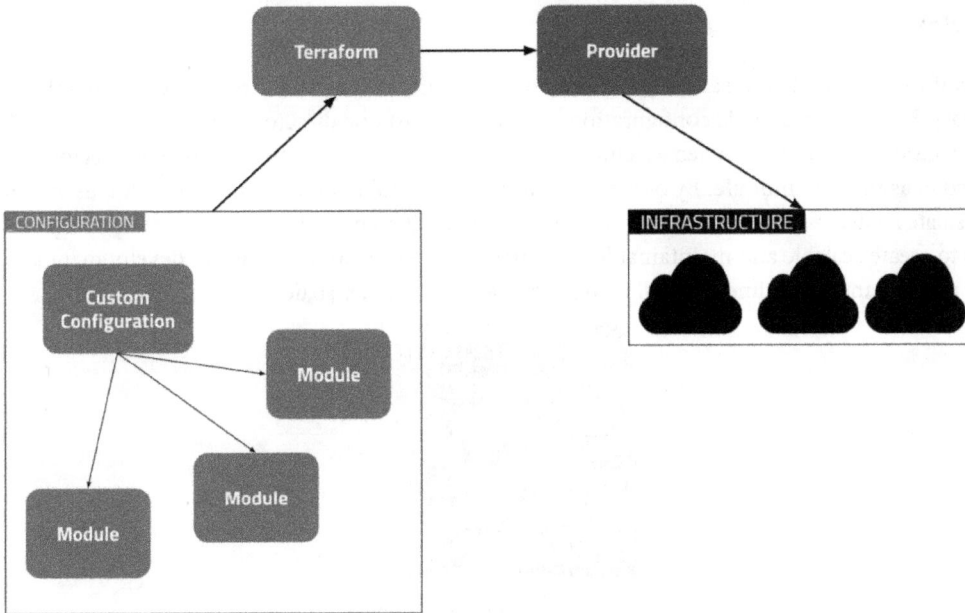

Figure 2.2 – Terraform technical flow

Providers

Terraform providers are essentially plugins that Terraform installs to interact with remote systems, such as Azure, AWS, Google Cloud, VMware, and a lot of other vendors' devices.

To interact with various cloud providers, SaaS providers, and other APIs, Terraform relies on plugins known as providers. These providers serve as the interface between Terraform and the target infrastructure, allowing Terraform to manage the desired resources. By leveraging a rich ecosystem of providers, Terraform enables users to manage infrastructure resources across multiple platforms using a unified, consistent syntax and workflow.

Terraform uses providers to provision resources, which describe one or more infrastructure objects such as virtual networks and compute instances. Each provider on the Terraform Registry has documentation detailing available resources and their configuration options.

If you haven't used any Terraform provider within your script, you won't be able to manage or create any infrastructure. You can define more than one provider in your Terraform code.

If you are working with Terraform modules, then you need to declare the Terraform providers in your root module while the child modules inherit the provider configuration from the root module. Terraform providers follow their own release cadence and version number syntax.

Modules

In Terraform, a module is essentially a collection of Terraform template files that reside in a single directory. Even the most basic configuration consisting of just one directory with one or more `.tf` files is considered a **module**. When executing Terraform commands directly from such a directory, it's referred to as the **root module**. By organizing infrastructure configurations into modules, users can encapsulate related resources and easily reuse them across different projects. This modularity enables teams to create scalable and maintainable infrastructure code, resulting in faster development and easier maintenance over time. The following shows a sample folder structure.

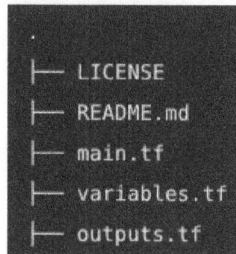

```
.
├── LICENSE
├── README.md
├── main.tf
├── variables.tf
├── outputs.tf
```

Figure 2.3 – Sample folder structure

In Terraform, commands operate solely on the template files located in one specific directory, typically the current working directory. However, by utilizing module blocks in your templates, you can invoke modules residing in other directories. Whenever Terraform encounters a module block, it processes the configuration files for that module.

When a template calls a module, the module is considered a "child module" of that template. Modules can be sourced either locally or remotely. Terraform supports multiple remote sources, such as the Terraform Registry, various version control systems, HTTP URLs, and private module registries in Terraform Cloud or Terraform Enterprise. By leveraging remote modules, teams can more easily share and collaborate on infrastructure configurations.

There are over 10,000 modules published in the public Terraform Registry ready to use at `https://registry.terraform.io/browse/modules`, and you can start to use them to provision infrastructure right away.

Let's go over the benefits of using modules:

- **Organize configuration**: Terraform offers modules as a means to simplify the organization and comprehension of your infrastructure configuration. This approach offers a more user-friendly experience, where users are not required to learn all of Terraform's features, but instead can focus on the specific features they need. Infrastructure can be complex, and even relatively simple deployments may require hundreds or thousands of lines of code. By leveraging modules, you can logically group related configuration elements into discrete components, making it easier to manage and modify your infrastructure over time.

- **Encapsulate configuration**: Modules offer the additional advantage of encapsulating configuration into distinct, logical components. This encapsulation can help avoid unintended consequences, such as inadvertently modifying other infrastructure while attempting to change a single component, and also minimize the likelihood of simple errors, such as inadvertently assigning the same name to different resources. By keeping related configuration elements organized within their own modules, users can more easily understand and modify their infrastructure in a controlled and systematic manner, reducing the risk of unexpected errors and making it simpler to implement changes over time.

- **Reuse configuration**: Creating infrastructure configurations from scratch can be a time-intensive and error-prone process. By utilizing modules, however, users can save time, enforce governance, and minimize costly errors by reusing pre-existing configurations authored by themselves, their colleagues, or other members of the Terraform community who have made modules publicly available. Additionally, modules can be shared within teams or published publicly, offering users the ability to benefit from the expertise of others and distribute their own work to others who may find it useful. Ultimately, leveraging modules can help users work more efficiently and effectively, making it easier to implement and maintain infrastructure configurations over time.

- **Provide consistency and ensure best practices**: In addition to facilitating efficient configuration development, modules also promote consistency across infrastructure configurations. Consistency is essential for comprehending complex configurations and ensures that best practices are applied uniformly across all configurations. For example, cloud providers offer numerous options for configuring object storage services such as Amazon S3 buckets. Improperly secured object storage has led to several critical security incidents, and due to the complexity of configuring these services, misconfiguration can occur easily.

- **Self-service**: Modules simplify the usage of your configuration by other teams. The Terraform Registry enables other teams to locate and utilize your approved and published Terraform modules.

- **Leveraging modules can enhance the governance of your resources**: For instance, modules can be utilized to define the configuration of your organization's public website buckets, as well as separate modules for private buckets that serve logging applications. Furthermore, modules facilitate updates to resource configurations by allowing you to modify a configuration in one location, which can then be applied to all instances where the module is used.

Resources

Terraform uses resource blocks to manage various kinds of infrastructure, such as virtual networks, compute instances, and higher-level components including DNS records. These resource blocks map to one or more infrastructure objects within your Terraform configuration.

Typically, each Terraform provider has several distinct resources that correspond to the relevant APIs for managing a given infrastructure type. Resource declarations can include several advanced features, although only a limited subset is necessary for initial use. With advanced syntax features, such as single resource declarations that generate multiple similar remote objects, users can familiarize themselves with and confirm all the features of a resource provider's documentation pages in the Terraform Registry.

Templates

Terraform templates provide a way to create resources in the desired format on the target cloud providers or systems.

A Terraform template is a collection of files that define the desired state of your infrastructure to be achieved. They include different configuration files, such as variables, resources, and modules. You can keep a single file or multiple files under a directory, depending on your needs and personal choice.

Now that we have covered the components of the architecture, let's look at the workflow.

Workflow

Terraform's workflows consist of five fundamental steps:

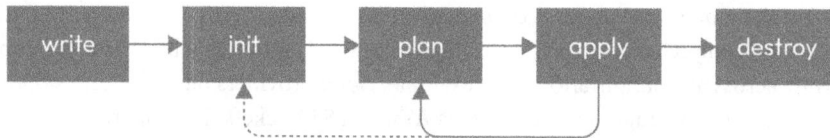

Figure 2.4 – Terraform's workflow

These steps involve the following:

1. **Write**: This step involves creating modifications to the code.
2. **Init**: At this stage, you initialize your code and download the requirements specified in your code, such as providers.
3. **Plan**: In this step, you review and predict the changes and determine whether to accept them.
4. **Apply**: This is where you accept the changes and implement them on real infrastructure.
5. **Destroy**: This final step involves destroying all the infrastructure you've created.

Details and actions vary between workflows. Let's look at all the steps of the workflow in detail.

Write

To begin with, author your Terraform configuration just like you would write code in your preferred editor. It is standard practice to store your work in a version-controlled repository, even when working individually.

Init

The `terraform init` command initializes the working directory where Terraform configuration files are located. It is recommended to execute this command as the first step after creating a new Terraform configuration or cloning an existing one from version control.

Executing this command multiple times is safe. `terraform init` carries out several initialization processes that prepare the current working directory for Terraform use. While subsequent runs may produce errors, the command will not delete your existing configuration or state.

Most providers are available as Terraform plugins. When executing the command, Terraform scans the configuration for direct and indirect references to providers and attempts to install the corresponding plugins. The program automatically discovers, downloads, and installs appropriate provider plugins that are published in either the public registry or a third-party provider's registry.

Terraform stores information about the chosen providers in the dependency lock file after a successful installation. To ensure that Terraform selects the same provider versions when you run `terraform init` in the future, commit this file to your version control system.

The `terraform init` command can be used for various purposes, such as plugin installation, child module installation, and backend initialization.

```
PROBLEMS    OUTPUT    DEBUG CONSOLE    TERMINAL

PS C:\Users\user\Desktop\Terraform\demo> terraform init

Initializing the backend...

Initializing provider plugins...
- Reusing previous version of hashicorp/aws from the dependency lock file
- Using previously-installed hashicorp/aws v4.14.0

Terraform has made some changes to the provider dependency selections recorded
in the .terraform.lock.hcl file. Review those changes and commit them to your
version control system if they represent changes you intended to make.

Terraform has been successfully initialized!

You may now begin working with Terraform. Try running "terraform plan" to see
any changes that are required for your infrastructure. All Terraform commands
should now work.

If you ever set or change modules or backend configuration for Terraform,
rerun this command to reinitialize your working directory. If you forget, other
commands will detect it and remind you to do so if necessary.
PS C:\Users\user\Desktop\Terraform\demo>
```

Figure 2.5 – Terraform initialization output

Plan

The `terraform plan` command generates an execution plan that enables you to preview the infrastructure modifications Terraform intends to make. Once a plan is generated, Terraform performs the following actions:

- Reads the current state of existing remote objects to verify that the state is current

- Compares the current state to the previous state, noting any discrepancies

- Suggests a set of actions that, if implemented, should ensure that the remote objects align with the configuration

The `terraform plan` command itself does not implement the predicted modifications. Instead, it is intended for use before applying or sharing your changes with your team to verify that the proposed changes align with your expectations. If Terraform detects no resource changes, the `terraform plan` command indicates that no changes are required for the actual infrastructure.

```
PROBLEMS    OUTPUT    DEBUG CONSOLE    TERMINAL

PS C:\Users\user\Desktop\Terraform\demo> terraform plan

Terraform used the selected providers to generate the following execution plan. Resource actions are indicated with the
following symbols:
  + create

Terraform will perform the following actions:

  # aws_instance.web will be created
  + resource "aws_instance" "web" {
      + ami                                  = "ami-065efef          b"
      + arn                                  = (known after apply)
      + associate_public_ip_address          = (known after apply)
      + availability_zone                    = (known after apply)
      + cpu_core_count                       = (known after apply)
      + cpu_threads_per_core                 = (known after apply)
      + disable_api_termination              = (known after apply)
      + ebs_optimized                        = (known after apply)
      + get_password_data                    = false
      + host_id                              = (known after apply)
      + id                                   = (known after apply)
      + instance_initiated_shutdown_behavior = (known after apply)
      + instance_state                       = (known after apply)
      + instance_type                        = "t2.micro"
      + ipv6_address_count                   = (known after apply)
      + ipv6_addresses                       = (known after apply)
      + key_name                             = (known after apply)
      + monitoring                           = (known after apply)
      + outpost_arn                          = (known after apply)
      + password_data                        = (known after apply)
      + placement_group                      = (known after apply)
      + placement_partition_number           = (known after apply)
      + primary_network_interface_id         = (known after apply)
      + private_dns                          = (known after apply)
      + private_ip                           = (known after apply)
```

Figure 2.6 – terraform plan output

Apply

To execute the actions suggested by Terraform plan, use the `terraform apply` command. The simplest approach is to execute `terraform apply` without any arguments, which automatically generates a new execution plan (similar to running `terraform plan`) and asks for approval before executing the proposed actions.

Unless explicitly directed to skip approval, `terraform apply` prompts the user for confirmation before making any changes to the corresponding infrastructure provider's API.

If no changes are detected in the configuration files compared to the current Terraform state, no modifications will be made to the infrastructure. Since Terraform is a declarative language, the `terraform apply` command can be executed multiple times safely.

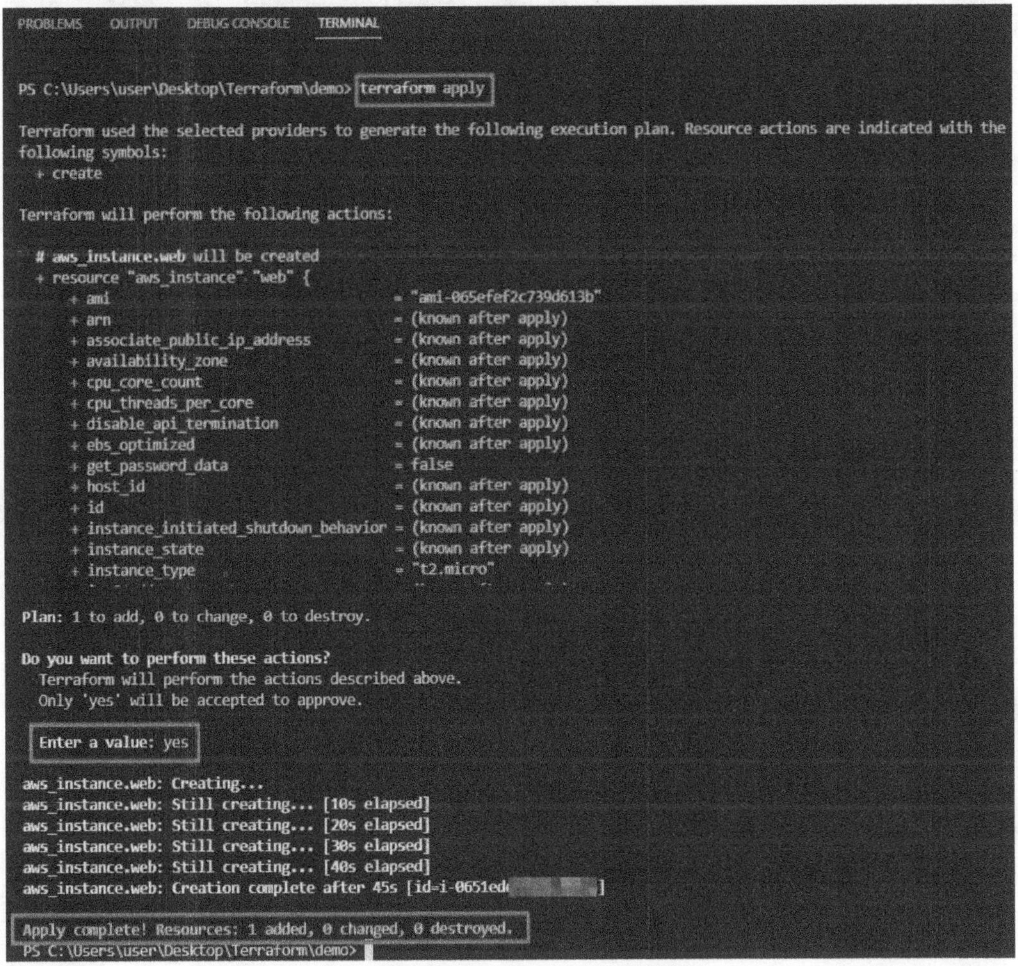

Figure 2.7 – terraform apply command

Destroy

You can use the `terraform destroy` command to easily destroy all remote objects managed by a specific Terraform configuration. You should not destroy objects that last for a long time in a production environment, but sometimes Terraform is used to handle short-lived infrastructure for development purposes, where `terraform destroy` can helpfully remove all temporary objects when you don't need them anymore.

The `terraform destroy` command should be used with caution and is not a command you would execute regularly. However, it is frequently used in non-production environments, where cleanup tasks are necessary for many proof-of-concept tests.

```
PROBLEMS    OUTPUT    DEBUG CONSOLE    TERMINAL

PS C:\Users\user\Desktop\Terraform\demo> terraform destroy
aws_instance.web: Refreshing state... [id=i-0651ede29        2a]

Terraform used the selected providers to generate the following execution plan. Resource actions are indicated with the
following symbols:
  - destroy

Terraform will perform the following actions:

  # aws_instance.web will be destroyed
  - resource "aws_instance" "web" {
      - ami                                  = "ami-065efef2c        3b" -> null
      - arn                                  = "arn:aws:ec2:us-east-1:7                                    d982a" -> null
      - associate_public_ip_address          = true -> null
      - availability_zone                    = "us-east-1b" -> null
      - cpu_core_count                       = 1 -> null
      - cpu_threads_per_core                 = 1 -> null
      - disable_api_termination              = false -> null
      - ebs_optimized                        = false -> null
      - get_password_data                    = false -> null
      - hibernation                          = false -> null
      - id                                   = "i-0651ede29        2a" -> null
      - instance_initiated_shutdown_behavior = "stop" -> null
      - instance_state                       = "running" -> null
      - instance_type                        = "t2.micro" -> null
      - ipv6_address_count                   = 0 -> null
      - ipv6_addresses                       = [] -> null
      - monitoring                           = false -> null
      - primary_network_interface_id         = "eni-0275306302414e5b5" -> null
      - private_dns                          = "ip-172-31-30-71.ec2.internal" -> null
      - private_ip                           = "172.31.30.71" -> null
      - public_dns                           = "ec2-18-209-50-142.compute-1.amazonaws.com" -> null
      - public_ip                            = "18.209.50.142" -> null

Plan: 0 to add, 0 to change, 1 to destroy.

Do you really want to destroy all resources?
  Terraform will destroy all your managed infrastructure, as shown above.
  There is no undo. Only 'yes' will be accepted to confirm.

  Enter a value: yes

aws_instance.web: Destroying... [id=i-0651ede293    82a]
aws_instance.web: Still destroying... [id=i-0651ede        a, 10s elapsed]
aws_instance.web: Still destroying... [id=i-0651ede        a, 20s elapsed]
aws_instance.web: Still destroying... [id=i-0651ede        a, 30s elapsed]
aws_instance.web: Destruction complete after 31s

Destroy complete! Resources: 1 destroyed.
PS C:\Users\user\Desktop\Terraform\demo>
```

Figure 2.8 – terraform destroy command

In summary, Terraform provides a set of commands to facilitate the creation, modification, and deletion of infrastructure resources. The `terraform init` command initializes the working directory where the Terraform configuration files are located. `terraform plan` generates an execution plan to preview the changes to be made to the infrastructure, while `terraform apply` executes the suggested changes after receiving user confirmation. Finally, `terraform destroy` destroys all remote objects managed by the configuration. With these commands, Terraform provides a powerful, flexible, and efficient toolset for managing IaC. Let's compare other IaC tools to Terraform.

To Compare with the Other IaC Tools

Terraform's flexible abstraction of resources and providers allows it to represent a wide range of infrastructure components, from physical hardware and virtual machines to email and DNS providers. This versatility makes Terraform capable of addressing various issues.

Terraform can manage virtually any cloud or virtual environment, including AWS, Microsoft Azure, and Google Cloud Platform, among others.

While this chapter focuses on using Terraform to manage AWS infrastructure, it is essential to note that Terraform is not exclusive to only cloud platforms. It can manage a single application or an entire data center.

Terraform versus CloudFormation

When it comes to **IaC** tools for cloud-based resources, two of the most popular choices are Terraform and AWS CloudFormation. While both tools aim to provide a reliable, efficient, and safe way to manage cloud infrastructure, they differ in their approach and implementation. Terraform is an open source tool that offers a flexible and extensible language for creating and managing infrastructure. In contrast, CloudFormation is an **AWS** proprietary tool that uses JSON or YAML templates to define infrastructure resources. This section will compare and contrast the features and capabilities of Terraform and CloudFormation, to help you make an informed decision on which tool best suits your infrastructure management needs.

What is AWS CloudFormation?

CloudFormation is a service provided by **AWS** that simplifies the management of a collection of AWS resources, including provisioning and updating them as necessary. With CloudFormation, you can create, update, and delete stacks in response to changes in your application environment. This AWS-managed service also offers a straightforward approach to creating reusable templates that enable you to deploy cost-effective applications.

CloudFormation allows you to design and provision AWS and third-party resources for your cloud environment using a configuration format known as templates. These templates, written in either JSON or YAML format, allow for the reusability and scalability of infrastructure, making it easier to manage large-scale cloud environments. The following diagram illustrates how Amazon CloudFormation serves as the central orchestrator for various AWS services.

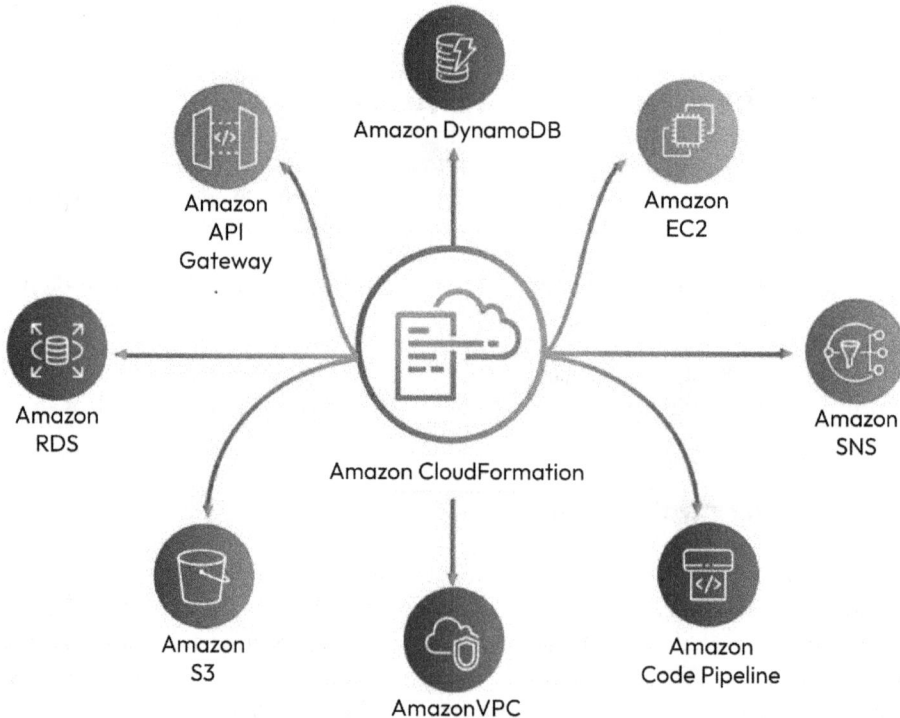

Figure 2.9 – AWS CloudFormation

Comparison and differences between Terraform and CloudFormation

Scope: On the coverage side, CloudFormation is very powerful because it is developed and supported directly by AWS, but Terraform has a great community that always works at a fast pace to ensure new resources, and features are implemented for providers quickly.

Type: CloudFormation is a managed service by AWS, but Terraform has a CLI tool that can run from your workstation, a server, or a CI/CD system (such as Jenkins, GitHub Actions, etc.) or Terraform Cloud (a SaaS automation solution from HashiCorp).

License and support: CloudFormation is a native AWS service, and AWS Support plans cover it as well. Terraform is an enterprise product and an open source project. HashiCorp offers 24/7 support, but at the same time, the huge Terraform community and provider developers are always helpful.

Syntax/language: CloudFormation supports both JSON and YAML formats. Terraform uses **HashiCorp Configuration Language (HCL)**, which is human-readable as well as machine-friendly.

Architecture: CloudFormation is an AWS-managed service to which you send/upload your templates for provisioning; on the other hand, Terraform is a decentralized system with which you can provision infrastructure from any workstation or server.

Modularization: In CloudFormation, nested stacks and cross-stack references can be used to achieve modularization, while Terraform is capable of creating reusable and reproducible modules.

User experience/ease of use: In contrast to CloudFormation, which is limited to AWS services, Terraform spans multiple cloud service providers such as AWS, Azure, and Google Cloud Platform, among others. This flexibility allows Terraform to provide a unified approach to managing cloud infrastructure across multiple providers, making it a popular choice for organizations that use more than one cloud provider.

Life cycle and state management: CloudFormation stores the state and manages it with the use of stacks. Terraform stores the state on disk in JSON format and allows you to use a remote state system, such as an AWS S3 bucket, that gives you the capability of tracking versions.

Import from existing infrastructure: It is possible to import resources into CloudFormation, but only a few resources are supported. It is possible to import all resources into Terraform state, but it does not generate configuration in the process; you need to handle that. But there are third-party tools that can generate configuration, too.

Verification steps: CloudFormation uses change sets to verify the required changes. Terraform has a powerful plan for identifying changes and allows you to verify your changes to existing infrastructure before applying them.

Rolling updates and rollbacks: CloudFormation automatically rolls back to the last working state. Terraform has no feature for rolling updates or rollbacks, but you can build a rollback system using a CI/CD system.

Multi-cloud management: CloudFormation is AWS-only, but Terraform supports multiple cloud providers and many more services.

Compliance integration: CloudFormation is built by AWS, so compliance is already assured, but for Terraform, you need to implement third-party tools yourself to achieve compliance.

Deployment type: CloudFormation has a built-in CI/CD system that takes care of everything concerning deployment and rollbacks. Terraform can be deployed from any system, but you need to build your CI/CD workflow or adopt a service that can fill the gaps.

Drift detection: Both tools have drift detection by default.

Cost: Using AWS CloudFormation does not incur any additional charges beyond the cost of the AWS resources that are created, such as Amazon EC2 instances or Elastic Load Balancing load balancers. In contrast, Terraform is an open source project that can be used free of charge. However, to obtain enterprise-level features such as CI/CD automation and state management, you may need to consider using additional services and systems provided by HashiCorp or third-party service providers. These additional services may come with their own costs.

Terraform or CloudFormation – which should I choose?

The debate between Terraform and CloudFormation is ongoing and ultimately, the decision of which tool to use will depend on your personal preferences and requirements. Both tools offer unique benefits and features, so it's important to evaluate which one aligns best with your organization's goals and cloud infrastructure needs.

CloudFormation is a more fitting choice when your whole infrastructure relies on AWS and you have no intention of integrating a multi-cloud setup in the future. For newcomers to AWS services, the native AWS provided integration support can be quite advantageous. As it's developed by AWS, CloudFormation enjoys quicker AWS-centric updates. It operates on JSON and YAML, thereby bypassing any potential learning hurdles associated with HCL.

On the other hand, Terraform shines when there's usage (even potential future usage) of multi-cloud resources and faster operation is desired. Its module-based design promotes the creation of repeatable templates, accelerating the configuration process. Additionally, Terraform provides a broader range of functions not found in CloudFormation, which is beneficial for quicker resource provisioning.

What is best for you ultimately depends on your requirements. You should assess your application's infrastructure strategy, security and compliance requirements, and cloud adoption strategy to make a final decision.

Summary

This chapter provided an overview of Terraform and its architecture and workflow, as well as a comparison between Terraform and CloudFormation. Terraform is a popular IaC tool that allows for safe and efficient management of infrastructure across multiple cloud providers. It relies on plugins called providers to interact with cloud providers, SaaS providers, and other APIs. Terraform uses modules to organize and encapsulate configurations into logical components, making it easier to navigate and understand complex configurations. Using modules can also provide consistency in your configurations and make them easier for other teams to use. Terraform uses resource blocks to manage infrastructure objects in your configuration, with most providers having several different resources that map to the appropriate APIs to manage that infrastructure type. The workflow of Terraform consists of five key steps: **Write**, **Init**, **Plan**, **Apply**, and **Destroy**.

The chapter also compared Terraform to AWS CloudFormation, which is limited to AWS services, whereas Terraform can span multiple cloud service providers. While CloudFormation simplifies the process of managing the life cycle of AWS resources and provides a simple way to create reusable templates, Terraform's flexibility allows for a unified approach to managing cloud infrastructure across multiple providers.

Ultimately, the decision of whether to use Terraform or CloudFormation will depend on personal preference and requirements. Both tools offer unique benefits and features that can help organizations efficiently manage their cloud infrastructure.

In the upcoming chapter, we will delve into various key aspects of working with Terraform. Let's start coding our first Terraform template.

3

Building Your First Terraform Project

If you're new to Terraform and looking to get started with your first project, this chapter is for you. In this chapter, we'll cover the basics of building your first Terraform configuration for AWS. We'll start by discussing how to install Terraform and prepare it for use with AWS. Then, we'll guide you through the process of building your first Terraform configuration and template. We'll also show you how to provision and test your first template so you can see your infrastructure come to life. By the end of this chapter, you'll have the foundational knowledge and skills to build your own infrastructure with Terraform.

In this chapter, we will cover the following topics:

- How to install Terraform
- How to install/prepare Terraform for AWS
- Building your first Terraform configuration
- Building your first Terraform template
- Provisioning and testing your template

How to install Terraform

To get started with Terraform, it's important to understand how to properly install and manage the Terraform installation. Installing Terraform can be challenging, but there are many online resources available to guide you through the process, including the official Terraform documentation. Terraform is distributed as a binary package by HashiCorp, and it can also be installed using popular package managers. Installing Terraform is the first step to creating your first project on Terraform.

Let's go over different installation methods next.

Manual installation

For manual installation, you have the choice of downloading a pre-compiled binary from the Terraform **Downloads** page, or compiling a binary from the source.

Pre-compiled binary

To install Terraform, you need to download the correct package for your operating system in the form of a ZIP archive. The appropriate package can be found by selecting your operating system on the Terraform website: `https://www.terraform.io/downloads`.

Once you have downloaded the appropriate Terraform package for your system, the next step is to unzip the package. Inside the package, you will find a single binary named `terraform`, which is the main executable for Terraform. You can safely remove any other files in the package, and Terraform will still function properly.

To use Terraform, you also need to ensure that the `terraform` binary is added to your system's `PATH` environment variable so that it can be executed from any directory in the terminal. The process for setting this up varies depending on your operating system, but instructions can typically be found online or in the Terraform documentation.

Mac or Linux PATH configuration

Print a colon-separated list of locations in your `PATH`:

```
echo $PATH
```

Move the Terraform binary to one of the listed locations. This command assumes that the binary is currently in your downloads folder and that your `PATH` includes `/usr/local/bin`, but you can customize it if your locations are different:

```
mv ~/Downloads/terraform /usr/local/bin/
```

Windows PATH configuration

The `PATH` configuration is stored in the registry, which you can edit through the following interface:

1. Go to **Control Panel | System | System settings | Environment variables**.
2. Scroll down through the system variables until you find `PATH`.

    Click **edit** and make the change accordingly.

    Be sure to include a semicolon at the end of the previous variable as these are used as delimiters.

3. Launch a new console for the settings to take effect.

Compiling from source

If you wish to compile the Terraform binary from source, you can clone the HashiCorp Terraform repository:

```
git clone https://github.com/hashicorp/terraform.git
```

Navigate to the new directory:

```
cd terraform
```

Then, compile the binary. This command will compile the binary and store it in `$GOPATH/bin/terraform`:

```
go install
```

After installing Terraform, it is important to ensure that the `terraform` binary is added to your system's `PATH` environment variable so that it can be executed from anywhere. This process may vary depending on your operating system, but typically involves adding the location of the `terraform` binary to the `PATH` variable using a command or by manually editing system files.

Popular package managers

There are several popular package managers that simplify the process of installing Terraform on different operating systems. These package managers allow you to manage and update multiple software packages from a single **Command-Line Interface (CLI)**. In this section, we will explore some of the most popular package managers for Terraform, including Chocolatey for Windows, Homebrew for macOS, and APT and Yum for Linux.

Homebrew on macOS

To install Terraform on Mac macOS, you can utilize the free and open source package management system for Mac macOS known as Homebrew. Before installing Terraform itself, you need to install the HashiCorp `tap`, which is a repository that contains all the Homebrew packages:

```
brew tap hashicorp/tap
```

Now, install Terraform with `hashicorp/tap/terraform`:

```
brew install hashicorp/tap/terraform
```

> **Note**
>
> This installs a signed binary and is automatically updated with every new official release.

To update to the latest version of Terraform, first update Homebrew:

```
brew update
```

Then, run the `upgrade` command to download and use the latest Terraform version:

```
brew upgrade hashicorp/tap/terraform
```

Chocolatey on Windows

To install Terraform on Windows using Chocolatey, you can use the CLI. Chocolatey is a free and open source package manager for Windows that simplifies the process of installing and managing software on your system:

```
choco install terraform
```

> **Note**
>
> Chocolatey and the Terraform package are *not* directly maintained by HashiCorp. The latest version of Terraform is always available for manual installation.

Linux

HashiCorp officially maintains and signs packages for Ubuntu/Debian, CentOS/RHEL, Fedora, and Amazon Linux distributions.

Ubuntu/Debian

Make sure that your system is up to date, and that you have the gnupg, `software-properties-common`, and `curl` packages installed. These packages are necessary to verify the GPG signature of HashiCorp and install its Debian package repository:

```
sudo apt-get update && sudo apt-get install -y gnupg software-
properties-common
```

Install the HashiCorp GPG key:

```
gpg --no-default-keyring \
    --keyring /usr/share/keyrings/hashicorp-archive-keyring.gpg \
    --fingerprint
```

Add the official HashiCorp repository to your system by using the appropriate command for your system. The command uses `lsb_release -cs` to find the distribution release codename for your current system, such as `buster`, `groovy`, or `sid`, and adds it to the repository file:

```
echo "deb [signed-by=/usr/share/keyrings/hashicorp-archive-keyring.
gpg] \
    https://apt.releases.hashicorp.com $(lsb_release -cs) main" | \
    sudo tee /etc/apt/sources.list.d/hashicorp.list
```

Download the package information from HashiCorp:

```
sudo apt update
```

Install Terraform from the new repository:

```
sudo apt-get install terraform
```

CentOS/RHEL

Install `yum-config-manager` to manage your repositories:

```
sudo yum install -y yum-utils
```

Use `yum-config-manager` to add the official HashiCorp Linux repository:

```
sudo yum-config-manager --add-repo https://rpm.releases.hashicorp.com/
RHEL/hashicorp.repo
```

Install Terraform from the new repository:

```
sudo yum -y install terraform
```

Fedora

Install `dnf config-manager` to manage your repositories:

```
sudo dnf install -y dnf-plugins-core
```

Use `dnf config-manager` to add the official HashiCorp Linux repository:

```
sudo dnf config-manager --add-repo https://rpm.releases.hashicorp.com/
fedora/hashicorp.repo
```

Install Terraform from the new repository:

```
sudo dnf -y install terraform
```

Amazon Linux

Install `yum-config-manager` to manage your repositories:

```
sudo yum install -y yum-utils
```

Use `yum-config-manager` to add the official HashiCorp Linux repository:

```
sudo yum-config-manager --add-repo https://rpm.releases.hashicorp.com/
AmazonLinux/hashicorp.repo
```

Install Terraform from the new repository:

```
sudo yum -y install terraform
```

Verifying the installation

Once you have completed the installation process, it is essential to verify that Terraform is installed correctly. To do this, you can open a new terminal session and list Terraform's available subcommands. This will help you ensure that Terraform is functioning correctly and that you can begin to create and manage infrastructure using Terraform:

```
terraform -help
```

If you are getting a response similar to the following, that means your installation was successful:

```
> terraform -help
Usage: terraform [-version] [-help] <command> [args]

The available commands for execution are listed below.
The most common, useful commands are shown first, followed by
less common or more advanced commands. If you're just getting
started with Terraform, stick with the common commands. For the
other commands, please read the help and docs before usage.
```

Now that we have covered how to install Terraform, let's move on to preparing Terraform for AWS by configuring our AWS account to allow Terraform to interact with the AWS services.

How to install/prepare Terraform for AWS

Once Terraform is installed, you can create your first infrastructure in AWS. Terraform generates an execution plan outlining the necessary steps to reach the desired infrastructure state and then executes it. As you make changes to the configuration, Terraform can determine the differences and create incremental execution plans to apply the changes.

In the following steps, let's create an S3 bucket in AWS. First of all, there are some prerequisites to provision AWS resources in Terraform.

Prerequisites

After a successful Terraform installation, you need the following ready for the next steps:

- The AWS CLI installed
- An AWS account and associated credentials with permissions to create resources

AWS CLI installation

The **AWS CLI** is a comprehensive tool that enables you to manage your AWS services from a single, unified interface. It simplifies the management of multiple AWS services and allows for automation through scripts. By using the AWS CLI, you can execute AWS commands directly from the command line, making it easier to manage your AWS resources efficiently.

Linux

To install `aws-cli`, your system must have the capability to extract or unzip the downloaded package. If your operating system does not have the built-in `unzip` command, you can use an equivalent. The AWS CLI requires `glibc`, `groff`, and `less`. These components are typically included by default in most major distributions of Linux.

Download the installation package as follows:

```
curl "https://awscli.amazonaws.com/awscli-exe-linux-x86_64.zip" -o
"awscliv2.zip"
```

Extract the installer by unzipping it. If your Linux distribution does not have a built-in `unzip` command, you can use an alternative to extract it. The following command is an example of unzipping the package and creating a directory called `aws` under the current directory:

```
unzip awscliv2.zip
```

Execute the installation command using the install file located in the extracted AWS directory. The default installation path is `/usr/local/aws-cli` and a symbolic link will be created in `/usr/local/bin`. The command may require the use of `sudo` to grant write permissions to those directories:

```
sudo ./aws/install
```

Confirm the installation with the following command:

```
aws –version
```

You should receive a response similar to the following:

```
aws-cli/2.7.24 Python/3.8.8 Linux/4.14.133-113.105.amzn2.x86_64
botocore/2.4.5
```

macOS

If you have sudo/admin permissions, you can install the AWS CLI for all users on the computer with the following commands.

Download the file using the curl command. The -o option specifies the filename that the downloaded package is written to. In this example, the file is written to AWSCLIV2.pkg in the current folder:

```
curl "https://awscli.amazonaws.com/AWSCLIV2.pkg" -o "AWSCLIV2.pkg"
```

This process is different on macOS systems, where you can run the standard macOS installer program to install Terraform. You simply need to specify the downloaded .pkg file as the source using the -pkg parameter to specify the package name and the -target / parameter to specify the installation drive. During the installation process, the files are installed to /usr/local/aws-cli, and a symlink is automatically created in /usr/local/bin. However, to grant write permissions to those folders, you must include sudo in the command:

```
sudo installer -pkg ./AWSCLIV2.pkg -target /
```

To verify that the shell can find and run the aws command in your $PATH, use the following command:

```
which aws
```

The response should be the following:

```
/usr/local/bin/aws
```

Verify the PATH configuration and version:

```
aws -version
```

The response should be like the following:

```
aws-cli/2.7.24 Python/3.8.8 Darwin/18.7.0 botocore/2.4.5
```

If the aws command cannot be found, you might need to restart your terminal and re-verify.

Windows

You require admin rights to install aws-cli on your Windows system.

Download and run the AWS CLI MSI installer for Windows (64-bit): https://awscli.amazonaws.com/AWSCLIV2.msi

Alternatively, you can run the `msiexec` command to run the MSI installer from Command Prompt:

```
msiexec.exe /i https://awscli.amazonaws.com/AWSCLIV2.msi
```

You can verify whether the AWS CLI is installed on your Windows machine by opening a Command Prompt window and running the `aws --version` command:

```
aws -version
```

This will display the installed version of the AWS CLI, confirming its successful installation on your system.

The response should be something like the following:

```
aws-cli/2.7.24 Python/3.8.8 Windows/10 exe/AMD64 prompt/off
```

If Windows is unable to find the program, you might need to close and reopen the Command Prompt window to refresh the path and re-verify.

AWS account

If you do not have an AWS account, you can create one with the following steps; if you do already have one, please skip the AWS credentials configuration part:

1. Open the AWS home page at `https://aws.amazon.com/`.

2. Choose **Create an AWS account**.

3. Ensure that you provide accurate account information, including your email address, and select **Continue**. Entering an incorrect email address can result in being unable to access your AWS account.

4. Choose **Personal** or **Professional**.

5. Enter your company or personal information.

6. Read and accept the AWS Customer Agreement.

7. Choose **Create Account** and **Continue**.

8. Provide your payment information on the **Payment Information** page and select **Verify and Add** to proceed. It's mandatory to provide a valid payment method to complete the sign-up process.

9. After completing the payment information, you will be asked to verify your phone number. You need to select the country or REGION code from the list and provide a phone number that you can access in the next few minutes for verification.

10. Enter the code displayed in the CAPTCHA, and then submit.

11. After the automated system contacts you, you need to enter the PIN you receive and then choose the **Continue** option.

12. On the **Select a Support Plan** page, choose one of the available AWS Support plans.

13. Finally, wait for your new account to be activated. This usually takes a few minutes but can take up to 24 hours.

14. Once your account is fully activated, AWS will send you a confirmation email. Be sure to check your email inbox and spam folder for the confirmation message. After receiving the email message, you will have full access to all AWS services.

AWS credentials

In order to interact with AWS via programmatic methods, it's required to supply your AWS access keys. These keys authenticate your identity when making programmatic requests. Access keys could be either temporary credentials that are valid for a short-term period or long-term credentials, such as those linked to an IAM user or the AWS account's root user.

In order to use Terraform with AWS, it is recommended to create a dedicated user with specific permissions and generate access keys for the AWS CLI and Terraform CLI to communicate and provision resources.

Creating an IAM user and credentials for Terraform

Creating an IAM user is a standard practice in AWS for security and resource access management. It is possible to create multiple IAM users and provide them with specific permissions to AWS resources. The process of creating an IAM user is straightforward, and is typically done when new team members join the company or when a new application needs access to AWS resources. For Terraform, you need to create an IAM user and assign appropriate permissions to provision resources as follows:

1. Sign in to the AWS Management Console and open the IAM console at `https://console.aws.amazon.com/iam/`.

2. In the left navigation pane, choose **Users** and then **Add users**.

3. Use `terraform` as the username for the new user. This is the sign-in name for AWS.

4. Select the type of access this user will have. You can select programmatic access and access to the AWS Management Console; I recommend just selecting programmatic access so you do not need to maintain a programmatic user's password life cycle.

5. Choose **Next: Permissions**.

6. On the **Set permissions** page, specify how you want to assign permissions to this set of new users. You can add the `terraform` user to a group, copy permissions from an existing user, or attach policies directly. If you are planning to provision all resources with this user, you can attach the **AdministratorAccess** policy to the user, but this will provide broad permission to all resources in the account, so make sure to keep your credentials secret.

7. Choose **Next: Tags**.

8. (Optional) On the **Tags** page, add metadata to the user by attaching tags as key-value pairs.

9. Choose **Next: Review**. Verify the user permissions to be added to the new user. When you are ready to proceed, choose **Create user**.

10. You can view the access keys for the user by selecting the **Show** option next to each password and access key. To save the access keys, select the **Download .csv** option and save the file to a secure location.

   Now, you have a user and credentials for the next steps.

Now that you have installed and prepared Terraform for AWS, it's time to start building your first Terraform configuration. With your AWS access keys and user permissions in place, you can now begin to write your Terraform code to provision your infrastructure in the cloud.

Building your first Terraform configuration

After installing Terraform and the AWS CLI, it is time to configure connectivity between them to be able to create resources with Terraform.

To authenticate the Terraform AWS provider with your IAM credentials, you need to set the AWS_ACCESS_KEY_ID environment variable in your terminal by adding your key after the = sign:

```
export AWS_ACCESS_KEY_ID=
```

Then add your secret key as follows, again after the = sign:

```
export AWS_SECRET_ACCESS_KEY=
```

You can verify your credentials and connectivity with the following command:

```
aws sts get-caller-identity
```

And you should receive a similar output to the following:

```
{
    "Account": "1234567890",
    "UserId": "ABCDEFGHJIKLM",
    "Arn": "arn:aws:iam:: 1234567890:user/erol_kavas"
}
```

Now that you have successfully built your first Terraform configuration, let's move on to building your first Terraform template.

Building your first Terraform template

We have created our AWS account and an IAM user and set up the necessary credentials for Terraform and the AWS CLI to communicate with AWS to provision infrastructure. Let's now start developing our first Terraform template.

A dedicated working directory is required for each Terraform configuration. To begin with, create a new directory for your first Terraform project. Any code editor or terminal can be used for this purpose, and we will provide the terminal commands for your convenience:

```
mkdir my-first-terraform-project
```

Change into the directory so that we can start to create files in it:

```
cd my-first-terraform-project
```

Create an empty file to define your infrastructure:

```
touch main.tf
```

Open the `main.tf` file in your preferred text editor and copy the following configuration into the file. Save the file once you have added the configuration:

```
terraform {
  required_providers {
    aws = {
      source  = "hashicorp/aws"
      version = "~> 4.0"
    }
  }
}


# Configure the AWS Provider
provider "aws" {
  region = "us-east-1"
}


resource "aws_vpc" "example" {
  cidr_block = "10.0.0.0/16"
}
```

Here we explicitly state that we are using the AWS provider plugin from Terraform and demanding a version greater than 4.0. We do not provide the access key/secret from the user in IAM that we created, because they were already imported into the terminal, and you should never, ever hardcode your credentials in Terraform templates! We also supply the REGION in which we want to make all changes. In this example, we are using us-east-1.

We then use the `aws_vpc` resource identifier to state that we are trying to bring up a VPC instance, followed by the `example` name identifier. (This name can be anything you like.)

We provide the `cidr_block` information to give the VPC the cidr (IP address space) information.

Now we will run the `terraform init` command in the directory where we created our `main.tf` file to download and initialize the appropriate provider plugins. In this case, we are downloading the AWS provider plugin we specified in our `main.tf` file:

```
terraform init
```

The output should be as follows:

```
Initializing the backend...

Initializing provider plugins...
- Finding hashicorp/aws versions matching "~> 4.0"...
- Installing hashicorp/aws v4.36.1...
- Installed hashicorp/aws v4.36.1 (signed by HashiCorp)

Terraform has created a lock file .terraform.lock.hcl to record the
provider
selections it made above. Include this file in your version control
repository
so that Terraform can guarantee to make the same selections by default
when
you run "terraform init" in the future.

Terraform has been successfully initialized!

You may now begin working with Terraform. Try running "terraform plan"
to see
any changes that are required for your infrastructure. All Terraform
commands
should now work.

If you ever set or change modules or backend configuration for
Terraform,
rerun this command to reinitialize your working directory. If you
forget, other
commands will detect it and remind you to do so if necessary.
```

The providers are now installed and we have the project initialized. Let's validate our template with the following command:

```
terraform validate
```

The output must be as follows:

```
Success! The configuration is valid.
```

After verifying our configuration, we can use `terraform fmt` to format our configuration files to use the correct format and style. In Terraform's newer versions, the introduction of new formatting rules in `terraform fmt` aren't considered a breaking change. However, our objective is to keep changes to a minimum for configurations that already conform to the style guides provided in the Terraform documentation.

Provisioning and testing your template

After the validation, let's run the `terraform plan` command. This will let us see what Terraform will do before we decide to apply it:

```
terraform plan
```

The output should be as follows:

```
Terraform used the selected providers to generate the following
execution plan. Resource actions are indicated with the following
symbols:
  + create

Terraform will perform the following actions:

  # aws_vpc.example will be created
  + resource "aws_vpc" "example" {
      + arn                                  = (known after apply)
      + cidr_block                           = "10.0.0.0/16"
      + default_network_acl_id               = (known after apply)
      + default_route_table_id               = (known after apply)
      + default_security_group_id            = (known after apply)
      + dhcp_options_id                      = (known after apply)
      + enable_classiclink                   = (known after apply)
      + enable_classiclink_dns_support       = (known after apply)
      + enable_dns_hostnames                 = (known after apply)
      + enable_dns_support                   = true
      + enable_network_address_usage_metrics = (known after apply)
      + id                                   = (known after apply)
      + instance_tenancy                     = "default"
      + ipv6_association_id                  = (known after apply)
      + ipv6_cidr_block                      = (known after apply)
      + ipv6_cidr_block_network_border_group = (known after apply)
```

```
        + main_route_table_id                = (known after apply)
        + owner_id                           = (known after apply)
        + tags_all                           = (known after apply)
    }


Plan: 1 to add, 0 to change, 0 to destroy.
```

```
Note: You didn't use the -out option to save this plan, so Terraform
can't guarantee to take exactly these actions if you run "terraform
apply" now.
```

Terraform provides a dependable and secure way to manage your infrastructure life cycle through the use of declarative configuration files. The largest hurdle encountered when handling **Infrastructure as Code (IaC)** is a phenomenon known as drift. Drift refers to the discrepancy between the actual state of your infrastructure and the state articulated in your configuration.

`terraform plan` is a very important command to detect drift in Terraform-managed resources – you must be able to understand every output and change in the plan's output, especially the line with the summary. In our example, this is as follows:

```
Plan: 1 to add, 0 to change, 0 to destroy.
```

After investigating the changes and verifying that they are what we planned and intended, let's proceed with `terraform apply`:

```
terraform apply
```

After running `apply`, Terraform will run another plan and ask us to verify and approve the predicted changes. The output should be similar to the following:

```
Terraform used the selected providers to generate the following
execution plan. Resource actions are indicated with the following
symbols:
  + create

Terraform will perform the following actions:

  # aws_vpc.example will be created
  + resource "aws_vpc" "example" {
      + arn                              = (known after apply)
      + cidr_block                       = "10.0.0.0/16"
      + default_network_acl_id           = (known after apply)
```

```
        + default_route_table_id                    = (known after apply)
        + default_security_group_id                 = (known after apply)
        + dhcp_options_id                           = (known after apply)
        + enable_classiclink                        = (known after apply)
        + enable_classiclink_dns_support            = (known after apply)
        + enable_dns_hostnames                      = (known after apply)
        + enable_dns_support                        = true
        + enable_network_address_usage_metrics = (known after apply)
        + id                                        = (known after apply)
        + instance_tenancy                          = "default"
        + ipv6_association_id                       = (known after apply)
        + ipv6_cidr_block                           = (known after apply)
        + ipv6_cidr_block_network_border_group = (known after apply)
        + main_route_table_id                       = (known after apply)
        + owner_id                                  = (known after apply)
        + tags_all                                  = (known after apply)
    }

Plan: 1 to add, 0 to change, 0 to destroy.

Do you want to perform these actions?
  Terraform will perform the actions described above.
  Only 'yes' will be accepted to approve.

  Enter a value:
```

You should carefully review the output, and if everything looks good, you can approve it by typing yes. Other than yes, infrastructure provisioning will be discarded, and apply will not provision anything in your environment.

After your approval, terraform-cli will start to deploy the resources you have requested, and the additional output should be similar to the following:

```
aws_vpc.example: Creating...
aws_vpc.example: Creation complete after 2s [id=vpc-xxxxxx]
Apply complete! Resources: 1 added, 0 changed, 0 destroyed.
```

The last line gives the summary of your terraform apply command, allowing you to easily see what was deployed, added, changed, or destroyed.

After verifying the resource from the AWS web console, you can destroy the example resources created in this exercise by running the terraform destroy command:

```
terraform destroy
```

When you use Terraform to manage your infrastructure, the `terraform destroy` command can be used to terminate resources created by your Terraform project. It is the opposite of the `terraform apply` command because it removes all the resources specified in your Terraform state. However, it's important to note that the `terraform destroy` command does not destroy resources that are running elsewhere and are not managed by the current Terraform project.

The output should look like this:

```
aws_vpc.example: Refreshing state... [id=vpc-xxxx]
Terraform used the selected providers to generate the following
execution plan. Resource actions are indicated with the following
symbols:
  - destroy


Terraform will perform the following actions:

  # aws_vpc.example will be destroyed
  - resource "aws_vpc" "example" {
      - arn                                 = "arn:aws:ec2:us-east-
1:xxxx:vpc/vpc-xxxx" -> null
      - assign_generated_ipv6_cidr_block    = false -> null
      - cidr_block                          = "10.0.0.0/16" -> null
      - default_network_acl_id              = "acl-0cbfcf0156e3eec97"
-> null
      - default_route_table_id              = "rtb-0933253f8baad1cb2"
-> null
      - default_security_group_id           = "sg-09aa1459d60ec7939"
-> null
      - dhcp_options_id                     = "dopt-26ad5a5f" -> null
      - enable_classiclink                  = false -> null
      - enable_classiclink_dns_support      = false -> null
      - enable_dns_hostnames                = false -> null
      - enable_dns_support                  = true -> null
      - enable_network_address_usage_metrics = false -> null
      - id                                  = "vpc-xxxx" -> null
      - instance_tenancy                    = "default" -> null
      - ipv6_netmask_length                 = 0 -> null
      - main_route_table_id                 = "rtb-xxxx" -> null
      - owner_id                            = "xx" -> null
      - tags                                = {} -> null
      - tags_all                            = {} -> null
    }


Plan: 0 to add, 0 to change, 1 to destroy.
```

```
Do you really want to destroy all resources?
   Terraform will destroy all your managed infrastructure, as shown
above.
   There is no undo. Only 'yes' will be accepted to confirm.

   Enter a value:
```

The - prefix in `terraform destroy` indicates that the instance will be destroyed. Similar to `apply`, Terraform shows its execution plan and waits for approval before making any changes. It's important to carefully review all changes and approve them, as there is no way to recover the resources after approval.

Answer `yes` to execute this plan and destroy the infrastructure:

```
aws_vpc.example: Destroying... [id=vpc-xxxx]
aws_vpc.example: Destruction complete after 1s
Destroy complete! Resources: 1 destroyed.
```

Similar to the `apply` command, Terraform will determine the order in which to destroy resources. If there are multiple resources with dependencies, Terraform will destroy them in the appropriate order with regard to those dependencies. In this case, Terraform identified a single VPC with no dependencies and destroyed it.

Summary

In this chapter, we learned how to install Terraform and prepare it for use with AWS. We covered various installation methods including downloading the binary, using a package manager, and compiling from source. Additionally, we discussed how to set up an AWS account and create an IAM user for Terraform. We then walked through the process of creating a directory for our first Terraform project, pasting in configuration code, and using the `terraform apply` command to provision resources. Finally, we learned how to use `terraform destroy` to tear down the resources created by our Terraform project. With the skills learned in this chapter, you should now be able to create and manage infrastructure on AWS using Terraform.

In the following chapter, we will explore the utilization of Terraform in **IaC** projects.

Discovering Best Practices for Terraform IaC Projects

As you begin to work with Terraform, it's important to understand the best practices for **Infrastructure As Code (IaC)** projects. In this chapter, we'll explore some key best practices for Terraform, including how to maintain, execute, and secure your IaC projects. We'll also look at ways to implement Terraform within your DevOps or cloud teams. By following these best practices, you'll be able to create efficient and reliable infrastructure deployments with Terraform.

We will look at these main topics in this chapter:

- How to maintain IaC projects with Terraform
- How to execute IaC projects with Terraform
- How to secure IaC projects with Terraform
- Implementing Terraform in DevOps or cloud teams

How to maintain IaC projects with Terraform

Maintaining IaC projects with Terraform is crucial to ensure their longevity, accuracy, and security. This involves managing the state, updates, changes, and versions of infrastructure configurations, among other tasks. In this chapter, we will discuss best practices for maintaining IaC projects with Terraform, covering topics such as managing state, employing version control, testing, and more. By embracing a well-defined standard module structure, you lay the foundation for streamlined development, enhanced collaboration, and the mastery of resource orchestration at scale.

Follow a standard module structure

Modules in Terraform offer a way of packaging and reusing resource configurations. They are essentially containers that group multiple resources that are used together. Modules consist of a collection of `.tf` and/or `.tfvars` files kept together in a directory. Elevate your deployment game by following these best practices, each a stepping stone toward resourceful and harmonious provisioning:

- Start each module with a `main.tf` file, which contains the resources by default.

- Include a `README.md` file in Markdown format in every module. This file should contain basic documentation about the module.

- Create an `examples/` folder in each module with a separate subdirectory for each example. For each example, include a detailed `README.md` file.

- Use descriptive names for resource files, such as `network.tf`, `instances.tf`, or `loadbalancer.tf`, to create logical groupings of resources.

- Avoid creating a separate file for each resource. Instead, group resources by their shared purpose. For example, combine `google_dns_managed_zone` and `google_dns_record_set` in `dns.tf`.

- The module's root directory should only contain Terraform (`.tf`) and repository metadata files (such as `README.md` and `CHANGELOG.md`).

- Place additional documentation in a `docs/` subdirectory.

Adopt a naming convention

Naming conventions in IaC are very important for sustainability and setting a common practice for naming resources in teams:

- Use nouns for resource names. For example, you could name them `aws_instance` or `google_storage_bucket`.

- Make resource names singular; for example, use `aws_instance` instead of `aws_instances`.

- Use meaningful names for resources of the same type to differentiate between them; for example, you might name two load balancers `primary` and `secondary`.

- Use nouns for data source names – for example, `aws_availability_zones` and `google_project`.

- In some cases, data sources can return a list and can be plural. For example, `aws_availability_zones` returns a list of Availability Zones.

- Name the resource `main` to simplify references to a resource that is the only one of its type in the module. An example is `aws_security_group_rule.main`.

It takes extra mental work to remember `aws_resource.my_special_resource.id` versus `aws_resource.main.id`.

- Attribute names in Terraform configuration blocks should use all lowercase letters and underscores to separate words.

- Consistency is important, so all configuration objects, including resource types and data source types, should also use underscores to delimit multiple words.

  This is not recommended:

  ```
  resource "azurerm_virtual_machine" "main-vm" {
    name = "main-vm"
  }
  ```

  This is recommended:

  ```
  resource "azurerm_virtual_machine" "main_vm" {
    name = "main-vm"
  }
  ```

- In the resource name, don't repeat the resource type. For example, the following is not recommended:

  ```
  resource "azurerm_virtual_machines" "main_virtual_machine" { … }
  ```

  However, this is recommended:

  ```
  resource "azurerm_virtual_machine" "main" { ... }
  ```

Use variables carefully

Discover how to wield the might of variables judiciously, shaping dynamic environments and empowering your deployments with adaptability and efficiency:

- All variables should be declared in the `variables.tf` file.

- Choose meaningful names for variables that accurately describe their purpose or use within the Terraform configuration:

  - If you have inputs, local variables, or outputs that represent numeric values, such as disk sizes or RAM size, make sure to include the appropriate units in their names, e.g., `ram_size_gb`.

  - Boolean variables should be named with positive and meaningful names to simplify the conditional logic. For instance, a variable indicating whether external access is enabled can be named `enable_external_access`.

- Include descriptions for variables to provide additional context for new developers that descriptive names may not convey. Descriptions are automatically included in a published module's auto-generated documentation.

- Provide data types for defined variables.

- If appropriate, provide default values for variables:

  - Provide default values for variables that have environment-independent values, such as disk size.

  - For variables that have values specific to a particular environment (e.g., `project_id`), it is recommended not to provide default values. This ensures that the calling module provides the necessary values and avoids unintentional misconfiguration.

- Use empty defaults for variables (such as empty strings or lists) only when leaving the variable empty is a valid preference that the underlying APIs don't reject. Always use them in the same order as in the following example:

```
variable "global_settings" {
description = "Setting read in from a global settings block"
type = map
default = {}
}
```

- Use variables only for values that need to be different for each instance or environment. Before exposing a variable, make sure that there is a specific need to change that variable. If the possibility of a variable being used is minimal, it's best to not expose it at all.

- Always use a plural name when defining a variable of a type map or list, as many values will potentially be read in.

Expose outputs

Exposing outputs in Terraform can help other modules reference useful values and reduce the amount of redundant code in your project. Here are some tips for exposing your outputs:

- Organize all outputs in an `outputs.tf` file

- Use meaningful descriptions for all outputs

- Document output descriptions in the `README.md` file

- Utilize tools such as `terraform-docs` to auto-generate output descriptions on commit

- Output all useful values that root modules might need to refer to or share

- For open source or heavily used modules, expose all outputs that have the potential for consumption

- Avoid passing outputs directly through input variables, as this prevents them from being properly added to the dependency graph

- To ensure that implicit dependencies are created, make sure that outputs reference attributes from resources
- Instead of referencing an input variable for an instance directly, pass the attribute through, as shown in the example provided next.

    This is recommended:

    ```
    output "name" {
      description = "Name of Virtual Machine"
      value       = azurerm_virtual_machine.main.name
    }
    ```

    This is not recommended:

    ```
    output "name" {
      description = "Name of Virtual Machine"
      value       = var.name
    }
    ```

> **Note**
>
> You can find more details on the best practices for using Terraform at `https://cloud.google.com/docs/terraform/best-practices-for-terraform`.

Use data sources

Data sources enable Terraform to use data defined externally, such as in another Terraform configuration, a separate tool, or a function. Here are some examples:

- Place data sources next to the resources that reference them
- Consider moving a large number of data sources to a dedicated `data.tf` file
- Use variable or resource interpolation to fetch data relative to the current environment

Leverage tfvars files

For anything that's not a secret, use `tfvars` files as much as possible for all your inputs and add them to your source control. That way, you can keep track of values and revert to a previous commit if you make a mistake, and you can see what's deployed at a glance. This should also be where most of your deployment changes happen.

Separate variables and inputs based on their functionality

In your `variables.tf` file and `terraform.tfvars`, use comments to separate them based on their function. This makes your code more readable and easy to change when working with it.

```
### terraform.tfvars ###

# Setup inputs
billing_account   = "00000-00000-00000"
org_id            = "1111111111"
logging_folder_id = "folders/222222222"

# Impersonation Inputs
automation_project_id = "whatever-thingy"
impersonate_sa_email  = "whatever-thingy-automation@blahbleblucordonblue.iam.gserviceaccount.com"

# Project Naming Inputs
prefix        = "cool"
project_label = "rad"
```

Figure 4.1 – An example tfvars

Limit the use of custom scripts

Limit the use of scripts in your Terraform configuration to only when it is necessary. Keep in mind that the state of resources created through scripts is not managed by Terraform and can cause issues in the future. It is recommended to use Terraform's built-in resource types and providers whenever possible. Here are some key steps:

- Try to avoid using custom scripts as much as possible. Instead, rely on Terraform resources to define the desired behavior for your infrastructure. However, if there are cases where Terraform resources don't provide the functionality you need, use custom scripts sparingly. Keep in mind that resources created through scripts are not accounted for or managed by Terraform, so they may introduce complexities in managing your infrastructure.

- Clearly document the reason for any custom scripts used and have a deprecation plan, if possible.

- Provisioners in Terraform can be used to call custom scripts, including the `local-exec` provisioner.

- Organize custom scripts that Terraform will call using provisioners in a `scripts/` directory.

Include helper scripts in a separate directory

Organizing helper scripts in a separate directory is a good practice for maintaining Terraform IaC projects. To form a good foundation for your IaC project, consider the following:

- Keep helper scripts that are not called by Terraform in a directory named `helpers/`
- Document all helper scripts in the `README.md` file by providing a description and example invocations
- For helper scripts that accept arguments, provide argument-checking and `--help` output to ensure that the script is being used correctly

Put static files in a separate directory

Amplify your AWS infrastructure orchestration prowess by embracing the power of organization. Learn how strategically separating static files into dedicated directories empowers efficient resource management and elevates the clarity of your Terraform-powered deployments:

- Create a `files/` directory for static files that Terraform references but doesn't execute, such as startup scripts loaded onto Compute Engine instances
- Place lengthy HereDocs in external files, separate from their HCL code, and reference them with the `file()` function
- Use the `.tftpl` file extension for files that are read in by using the Terraform `templatefile()` function
- Place templates in a `templates/` directory

Protect stateful resources

Explore essential strategies to protect the core of your infrastructure, ensuring resilience, compliance, and seamless continuity.

Make sure to enable deletion protection for stateful resources such as databases. This is an example of enabling deletion protection:

```
resource "azurerm_sql_database" "main" {
  name = "primary-instance"
  settings {
    tier = " "
  }

  lifecycle {
    prevent_destroy = true
```

```
      }
    }
```

Use built-in formatting

It is important to maintain consistency and readability in Terraform code. To ensure this, use the Terraform `fmt` command to automatically format all Terraform files according to the official Terraform style guide. This helps maintain consistency and makes it easier for others to read and understand your code.

Limit the complexity of expressions

Avoid overly complex interpolated expressions; if an expression requires multiple functions, consider splitting it into smaller, more readable expressions using local values.

Use only one ternary operation per line, and use multiple local values instead of building up complex logic in a single line.

Use count for conditional values

Use the count meta-argument to conditionally instantiate a resource. This is an example:

```
variable "readers" {
  description = "..."
  type        = list
  default     = []
}
```

This Terraform code snippet declares a variable named `readers` with a list type and an empty default value. The description of the variable is not included in the code snippet:

```
resource "resource_type" "reference_name" {
  // Do not create this resource if the list of readers is empty.
  count = length(var.readers) == 0 ? 0 : 1
  ...
}
```

The code also defines a resource of type `resource_type` with the `reference_name` reference name. The count meta-argument is used to conditionally create the resource based on the length of the `readers` list variable. If the length of the `readers` list is 0, the resource will not be created (count = 0). Otherwise, if the length is greater than 0, the resource will be created (count = 1).

Use for_each for iterated resources

Use the `for_each` meta-argument when creating multiple copies of a resource based on an input resource.

Publish modules to a registry

Publish reusable modules to a module registry to increase reusability and make it easier for your team to use.

Next, we will explore how to secure IaC projects with Terraform.

How to execute IaC projects with Terraform

Every team always starts their Terraform journey by running Terraform from their development environments or local computers.

As the team starts to adopt more Terraform and IaC, you will need more automation to ensure consistency between runs and to provide other important features such as integration with version control, code reviews, environment management, etc.

Terraform automation can be achieved in diverse forms and to different extents. Some teams might persist in running Terraform locally, using customized wrapper scripts to ensure a uniform working directory for Terraform's operation. Meanwhile, other teams fully operate Terraform within orchestration tools such as Jenkins, GitHub Actions, Terraform Cloud, or Terraform Enterprise.

The following are steps to create automated execution pipelines for Terraform:

1. Select a version control system such as GitHub, Git, and so on.
2. Store all your templates/files in version control.
3. Select an automation pipeline.
4. Decide where to store your state (e.g., AWS S3, Terraform Cloud, etc.).
5. In the automation pipeline script, do the following:

    A. Verify that the Terraform **command-line interface (CLI)** exists in the automation pipeline.
    A. Verify connectivity to your state location and cloud provider.
    B. Initialize the Terraform working directory.
    C. Produce a plan for changing resources to match the current configuration.
    D. Have a human operator review that plan to ensure it is acceptable.
    E. Apply the changes described in the plan.

Having established the foundational aspects of executing IaC projects with Terraform, let's now shift our focus toward a critical facet: securing these projects. By seamlessly integrating robust security practices into your Terraform workflows, you not only safeguard your infrastructure but also fortify your development pipeline, ensuring the holistic success of your AWS environment.

How to secure IaC projects with Terraform

Using IaC or Terraform to deploy and manage resources makes the process faster and easier, eliminating the need for one-time scripts or manual steps. With Terraform, infrastructure can be managed in a similar way as applications and services, including servers, databases, networks, Kubernetes clusters, and entire application stacks.

While IaC may not present an immediate risk or attack surface, it's still important to consider security. However, because IaC is often managed by engineering and DevOps teams, security measures may be overlooked in favor of monitoring cloud resources already in production.

Managing infrastructure at scale can be complex, and security and DevOps teams may not have the necessary expertise, access, or tools to properly address security concerns. This can lead to misconfigured cloud resources, such as engineers and developers missing important security measures. Here are some common mistakes:

- Default configurations that haven't been optimized for security are used

- Logging is not enabled, making it difficult to troubleshoot or assemble an audit trail

- Unencrypted databases are used, leaving data vulnerable to corruption and exfiltration

- Insecure protocols (e.g., not using HTTPS) are deployed

- Foundational governance and prevention systems for security or misconfiguration are lacking

IaC involves defining and managing configurations using code, which means that all security configurations must also be defined in code. Here are some tips for securing your Terraform projects:

- Ensure that your state location is secure and not publicly accessible and provide appropriate access to your team for operations.

- Keep private modules in private module registries.

- Do not hardcode sensitive information such as secrets, credentials, keys, or certificates in Terraform files or variables. Instead, retrieve them from secure locations at runtime.

- Use a tool such as Checkov to scan your Terraform templates and directories for misconfigurations related to encryption, network, backup, IAM, and other security and compliance policies.

- Automate IaC scanning in your **continuous integration/continuous deployment (CI/CD)** pipeline for consistency and provide automated feedback as part of the CI run to prevent misconfigured code.

- Use a version control system, lock your main deployment branch, and always run automated checks and get peer reviews and approval before deployment.

- Carefully review the `terraform plan` results to avoid any unexpected resource destruction.

Having fortified your IaC projects with Terraform's security measures, the next significant stride is seamlessly integrating Terraform into your DevOps or cloud teams. Discover the pivotal steps to harmoniously weave Terraform into your team's practices, enhancing collaboration, efficiency, and the realization of a cohesive and empowered AWS environment.

Implementing Terraform in DevOps or cloud teams

Implementing Terraform in DevOps or cloud teams involves more than just adopting the tool. It also requires understanding the process, the team's capabilities, and the organization's culture. Here are some steps you can follow to successfully implement Terraform in your DevOps or cloud teams:

1. **Start with a pilot project**: Begin with a small project that can demonstrate the value of Terraform to the team. It could be a simple infrastructure deployment, such as a VPC, or a more complex application stack.

2. **Identify the team's knowledge gaps**: Terraform requires knowledge of cloud infrastructure, coding, and best practices. Identify any gaps in the team's knowledge and create a plan to address them.

3. **Encourage collaboration and knowledge sharing**: Terraform is a collaborative tool that requires contributions from multiple teams. Encourage your teams to share their knowledge, experiences, and best practices.

4. **Implement a version control system (VCS)**: Terraform code is code and should be treated like any other code. Implement a **VCS** to manage changes to the infrastructure code.

5. **Implement a CI/CD pipeline**: Automate the testing, building, and deployment of Terraform code using a CI/CD pipeline. This ensures that infrastructure changes are thoroughly tested before deployment.

6. **Implement IaC best practices**: Use best practices for IaC development, such as modularization, code reviews, and testing. This ensures that your infrastructure code is maintainable, scalable, and secure.

7. **Monitor and optimize**: Implement monitoring and alerting to identify and address issues as they arise. Optimize your infrastructure by reviewing logs and metrics and identifying areas for improvement.

By following these steps, you can successfully implement Terraform in your DevOps or cloud teams and reap the benefits of IaC.

Summary

In this chapter, we covered the best practices for Terraform IaC projects, including organizing files, defining variables, using data sources, and managing state. We also discussed how to maintain IaC projects with Terraform, including managing resources, deploying changes, and employing version control. Additionally, we explored how to execute IaC projects with Terraform, including using modules, providers, and Terraform Cloud.

Finally, we addressed how to secure IaC projects with Terraform, including securing state, avoiding hardcoded secrets, and using tools such as Checkov. By following these best practices, DevOps and cloud teams can effectively use Terraform to manage infrastructure at scale while maintaining security and compliance.

In the next chapter, we'll delve into the fundamentals of planning Terraform infrastructure projects. We'll kick off with guidance on crafting your initial Terraform template in AWS. Next, we'll deepen your comprehension of AWS providers and Terraform modules, essential elements for any Terraform project.

Part 2:
Become an Expert
in Terraform with AWS

In this section, we dive deeper into the intricacies of Terraform on AWS, guiding you through the process of becoming an expert in deploying and managing cloud infrastructure. We start by discussing the importance of planning and designing infrastructure projects in AWS, ensuring a solid foundation for your Terraform deployments. You'll learn how to make informed decisions for your AWS Terraform projects, considering factors such as resource selection, configuration, and deployment strategies. We then move on to the practical implementation of Terraform in various projects, including deploying serverless applications and containers in AWS. By the end of this part, you'll have a comprehensive understanding of how to expertly use Terraform to deploy and manage complex infrastructure on AWS.

This part contains the following chapters:

- *Chapter 5, Planning and Designing Infrastructure Projects in AWS*
- *Chapter 6, Making Decisions for Terraform Projects with AWS*
- *Chapter 7, Implementing Terraform in Projects*
- *Chapter 8, Deploying Serverless Projects with Terraform*
- *Chapter 9, Deploying Containers in AWS with Terraform*

5

Planning and Designing Infrastructure Projects in AWS

In the world of cloud computing, planning and designing infrastructure projects are crucial steps toward achieving the desired outcome. As the cloud computing environment evolves, it becomes increasingly important to have a proper plan and design in place for your infrastructure. With the help of Terraform, an infrastructure-as-code tool, you can easily plan and design your infrastructure projects in **Amazon Web Services** (**AWS**).

This chapter will provide you with the necessary knowledge to get started with planning and designing infrastructure projects using Terraform in AWS. We will discuss the basics of infrastructure project planning, designing your first Terraform template in AWS, understanding AWS Providers and Terraform modules, and implementing best practices with Terraform AWS modules. By the end of this chapter, you will have a solid foundation for planning and designing your infrastructure projects in AWS using Terraform.

In this chapter, we will cover these topics:

- Terraform infrastructure project planning basics

- How to design your first Terraform template in AWS

- Understanding AWS Providers

- Understanding Terraform modules

- How to implement best practices with Terraform AWS modules

Terraform infrastructure project planning basics

As technology advances, more businesses and leaders are beginning to adopt **infrastructure as code (IaC)** to manage their IT infrastructure. With the increasing need for flexibility and security in software development, IaC offers a high-level code solution to automate the provisioning of infrastructure resources. However, it's important to understand both the potential benefits and challenges that come with implementing IaC. In this chapter, we will explore the basics of planning and designing infrastructure projects in AWS using Terraform, covering important topics such as AWS Providers, Terraform modules, and best practices.

The speed benefits

The adoption of cloud computing has brought about IaC, which provides significant speed and agility benefits for deploying, modifying, and removing virtual infrastructure services. With IaC, teams can interact with infrastructure in a programmatic way, allowing for the automation of lifecycle management. In some cases, automation solutions can also manage non-programmatic, command-line-interface-based devices using IaC management.

The risk management benefits

Implementing IaC in an organization can offer numerous benefits. One of the primary advantages is the elimination of human error that often occurs during manual infrastructure provisioning and configuration. By using IaC, an organization can greatly minimize the risks associated with human error and enhance the security of its infrastructure.

Security, reusability, and governance

Setting up an IaC pipeline is crucial for organizations seeking to realize the full potential of this technology. Proper setup of an IaC pipeline requires considering factors such as security, reusability, and governance. It's essential to implement a complete continuous integration and continuous deployment pipeline that includes IaC, especially for applications that require frequent updates. This approach can significantly improve an organization's speed to market while reducing costs.

Team skill sets

When transitioning to an IaC platform, it's essential to consider the skill sets of the existing staff. IaC requires a different set of development skills that may not be present in the current team. Neglecting to take this into account can result in demotivation and disengagement among the staff, particularly if coding isn't a skill set that they have or are interested in acquiring. Therefore, it's essential to provide training and support to help employees adapt to this new way of working.

The best candidates for automation

Determining which infrastructure should be treated as code is a crucial decision. It's not worthwhile to automate infrastructure that will be deployed only once in the organization's lifetime, but it is worth automating infrastructure that will be regularly deployed for new applications or services. It's important not to get bogged down in automating everything; instead, make sure your IaC efforts provide a return on investment over traditional approaches.

The types of applications you'll be running

Designing the IaC template for an application is critical for its success. The configuration should be modularized and driven by the configuration management system, and the design process should take into account the application that will run on the infrastructure. For instance, if you are planning to deploy a database, the IaC design considerations will differ based on whether the application is transactional or for reporting purposes.

The cost of automating too many tasks

In the quest to improve efficiency, the general guideline is to automate any recurring tasks and reserve manual management for exceptional cases. This approach will help manage the expectations of internal stakeholders and departments effectively. However, it's essential to keep a close eye on the **return on investment (ROI)** for automation, as automating every infrastructure task can lead to cost overruns.

The critical nature of the code

As infrastructure is critical to the success of any business, the code used to manage it should be treated with the same level of importance. This includes having the right processes and backup procedures in place for when issues arise. Virtual networks, data centers, and servers require a disciplined approach to change management and testing, just like physical infrastructure.

The need for software expertise

Collaboration between software engineers and infrastructure engineers is crucial for successful IaC implementation. While infrastructure engineers are experts in managing and deploying infrastructure, they may lack knowledge of software development best practices. By embedding software engineers with the infrastructure team, organizations can bridge this gap and leverage the expertise of both teams to optimize results. Adopting an inner source model that encourages sharing and collaboration can further support this effort.

The impact on agility

Start-ups in the rapid growth stage may not be able to prioritize implementing IaC, as it can result in a lack of agility. While IaC brings many benefits to larger organizations, smaller companies need to balance implementing the necessary IaC and keeping their engineers motivated to think outside the box. As a technology company, it is important to maintain innovation and original ideas, which can be hindered by overreliance on IaC.

Integration with existing infrastructure

Before adopting IaC, businesses must assess the potential risks and benefits. Implementing IaC may come with adoption, security, and scalability challenges, such as integrating new frameworks with existing infrastructure. It requires a considerable amount of planning, time, and collaboration with other teams, including those responsible for security and compliance.

Goals and available resources

Introducing IaC into an organization requires careful planning and consideration of the goals for the transition. While IaC can bring significant benefits, such as reduced human error and increased security, there are also potential adoption, security, and scalability gaps to consider. It's essential to have a clear plan for integrating new frameworks with existing infrastructure and collaborating with other teams, including security and compliance.

Moreover, it's crucial to be mindful of the impact on existing tech and personnel resources. Pushing inexperienced engineers in a new direction can create an unstable organization. To minimize this risk, IaC should be incorporated as part of modernization efforts, with a focus on upskilling engineers to handle advanced projects. By doing so, organizations can ensure a smoother and more effective transition to IaC while maximizing its potential benefits.

The long-term plan

When implementing IaC, it's important to plan for the long term. This includes considering factors such as maintenance, security, and development time. It's also important to have an exit plan in place, which may involve multiple paths depending on various scenarios. By having a solid plan in place, you can ensure that your investment in IaC will pay off and that you can adapt to any changes or challenges that arise.

Quality control and security

Implementing IaC in the eDiscovery space requires a thoughtful and deliberate approach to avoid introducing unintended vulnerabilities. While traditional infrastructure deployment plans account for security holes, an IaC approach offers many benefits. However, to fully realize these benefits, it is essential to establish a comprehensive program that includes quality control and security measures. This will help ensure the stability and security of the infrastructure.

How to design your first Terraform template in AWS

Designing your first Terraform template in AWS can be an intimidating task if you're new to IaC. However, understanding the fundamentals and best practices can make the process much smoother. In this section, we will explore the key components of designing a Terraform template, including defining resources, understanding AWS Providers, and utilizing Terraform modules. We will also cover tips for implementing best practices with Terraform AWS modules, so you can design and deploy reliable and scalable infrastructure in AWS with confidence.

Authentication with AWS

To create and manage resources in AWS using Terraform, you first need to establish a connection between Terraform and AWS. This connection is authenticated using programmatic API keys, which consist of an access key and a secret key. These keys are used to access and manage your AWS resources through Terraform. In this section, we will explore some sample configurations that illustrate how to use API keys to provision your first infrastructure with Terraform in AWS:

```
provider "aws" {
  region     = "us-west-2"
  access_key = "my-access-key"
  secret_key = "my-secret-key"
}
```

To start, we should go and create these access and secret keys for Terraform from your AWS account.

Setting up programmatic access

Log in to the AWS Management Console, and then in the services, go to IAM, and perform the following steps:

1.   Add the new user and key in the **User name** field:

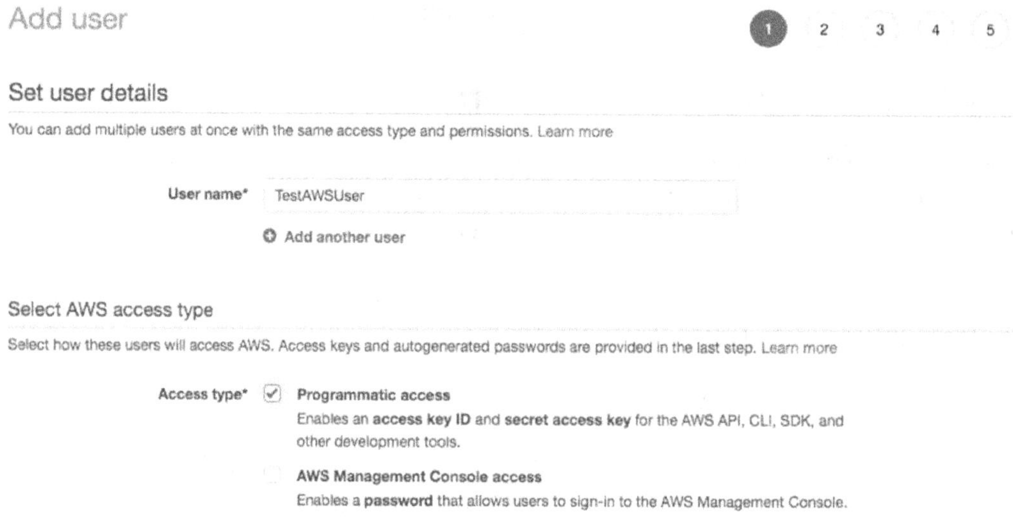

Add user   ① 2 3 4 5

Set user details

You can add multiple users at once with the same access type and permissions. Learn more

User name*   TestAWSUser

⊕ Add another user

Select AWS access type

Select how these users will access AWS. Access keys and autogenerated passwords are provided in the last step. Learn more

Access type*   ☑ **Programmatic access**
Enables an **access key ID** and **secret access key** for the AWS API, CLI, SDK, and other development tools.

AWS Management Console access
Enables a **password** that allows users to sign-in to the AWS Management Console.

Figure 5.1 – Adding user

2.   Select **Attach existing policies directly** and **AdmininistratorAccess**:

Add user   1 ② 3 4 5

▾ Set permissions

👥 Add user to group	👤 Copy permissions from existing user	🗎 Attach existing policies directly

Create policy   ⟳

Filter policies ⌄   Q Search   Showing 441 results

	Policy name ▾	Type	Used as	Description
☑ ▸	📖 AdmininistratorAccess	Job function	None	Provides full access to AWS services and re...

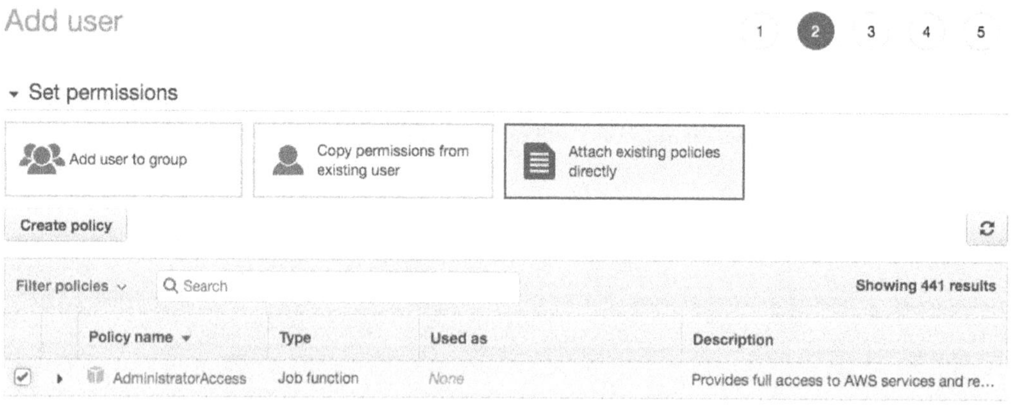

Figure 5.2 – Setting permissions

Click **Next** until you see the following screen.

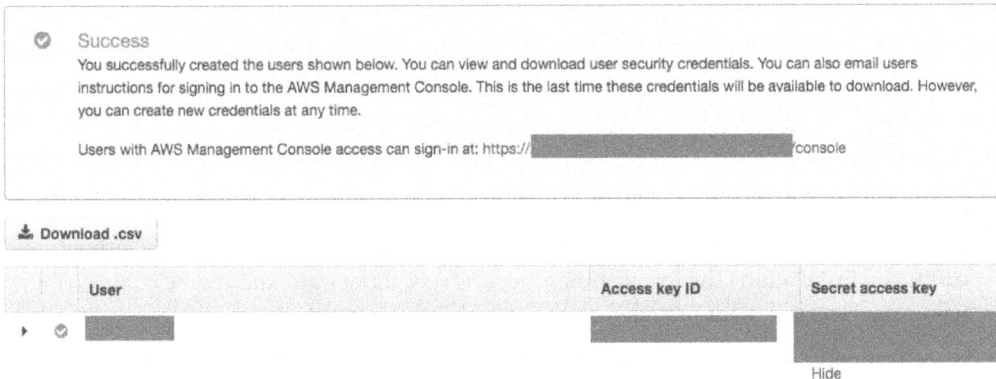

Figure 5.3 – Success screen

3. Complete the process and get your keys.

    After generating your access key ID and secret access key in the AWS console, it is important to securely store these credentials. While Terraform allows for the access key and secret key to be hardcoded within the configuration file, this approach is not recommended due to security risks. Instead, it is advised to save these keys as environment variables or as an AWS config profile.

    • Set as environment variables:

    To authenticate with AWS in your terminal or command line, you will need to run specific commands with your access and secret key.

    ```
    export AWS_ACCESS_KEY_ID=AK************IEVXQ
    export AWS_SECRET_ACCESS_KEY=gbaIbK********************iwN0dGfS
    ```

    • Set up as an AWS config profile:

    To get started, you'll need to download and set up the AWS CLI.

    Refer to this guide for instructions on how to install and configure the AWS CLI on your system:

    https://docs.aws.amazon.com/cli/latest/userguide/getting-started-install.html

    After successfully installing AWS CLI, here are the steps for configuring your profile that you can do from your terminal:

    First, you need to run the following:

    ```
    aws configure
    ```

After this, you will be asked to fill in the following information that you have downloaded from the AWS console:

```
AWS Access Key ID [None]: downloaded access key id
Secret Access Key [None]: downloaded secret access key
Default region name [None]: us-west-2
Default output format [None]: json
```

4.   Download and install Terraform CLI

To get started with Terraform, you can download the single-file binary and run it without any additional installation. The installation process is straightforward and can be completed by following the instructions provided on the official Terraform website.

Once Terraform is installed, you can start creating your IaC using the Terraform CLI:

```
https://developer.hashicorp.com/terraform/tutorials/aws-get-started/install-cli
```

5.   Terraform configuration:

Terraform requires a specific file known as the Terraform configuration file as input. This file is written in **HashiCorp Configuration Language** (**HCL**), but can also be written in JSON.

Here is a sample Terraform configuration file with the *.tf extension. This example assumes that the AWS config profile is being used and references the default profile for authentication:

```
provider "aws" {
  profile    = "default"
  region     = "us-east-1"
}
resource "aws_instance" "example" {
  ami            = "ami-2757f631"
  instance_type = "t2.micro"
}
```

If you are using the environment variables method for authentication, you can remove the profile line from the Provider block in your Terraform configuration file.

A Terraform configuration file consists of several elements, known as blocks, including Providers, resources, and more.

Here is an example of how the syntax for a Terraform configuration file block is formatted:

```
resource "aws_vpc" "main" {
  cidr_block = var.base_cidr_block
}


<BLOCK TYPE> "<BLOCK NAME>" "<BLOCK LABEL>" {
  # Block body
```

```
         <IDENTIFIER> = <EXPRESSION> # Argument
      }
```

Terraform offers a wide range of BLOCK_TYPE options, with the primary one being the resource. The other blocks support building the specified resource. These blocks include providers, which represent providers such as AWS, Google, and Azure:

- providers: Specifies the name of the Provider, such as AWS, Google, and Azure
- resources: Specifies a specific resource within the Provider, such as aws_instance for AWS
- variable: Declares input variables for the Terraform configuration
- output: Declares output variables that will be stored in the Terraform state file
- local: Assigns a value to an expression, which can be used as a temporary variable within a module
- module: A container for multiple resources that are used together
- data: Collects data from the remote Provider and saves it as a data source

Create your first AWS infrastructure with Terraform

Here are the steps to practically apply Terraform and create an EC2 instance:

1. Create a directory for your Terraform project and save the following code as a file named main. tf.
2. Initialize the directory using the terraform init command.
3. Verify the proposed changes by running the terraform plan command.
4. If you are satisfied with the changes that Terraform plans to make, execute terraform apply to commit and provision the AWS infrastructure.

Step 1 – creating a template file for Terraform AWS infrastructure

To create an EC2 instance in AWS using Terraform, we first need to create a directory and generate a Terraform configuration file named main.tf. It's important to ensure that no other *.tf files are present in the directory, as Terraform considers all files ending with the .tf extension as part of the provisioning process.

We can copy the following content and save it as main.tf:

```
provider "aws" {
   profile    = "default"
   region     = "us-east-2"
}
resource "aws_instance" "example" {
```

```
    ami          = "ami-0fb653ca2d3203ac1"
    instance_type = "t2.micro"
}
```

To provision an AWS EC2 instance using Terraform, you need to set the required arguments for the `aws_instance` resource. While there are many different arguments available, for this example, we will only set the two required arguments:

- `ami`: To launch an EC2 instance with Terraform, you need to specify the **Amazon Machine Image** (**AMI**) to run on the instance. You can find free and paid AMIs in the AWS Marketplace or create your own using tools such as Packer. In the code provided, the `ami` parameter is set to the ID of an Ubuntu 20.04 AMI in the `us-east-2` region, which is free to use. It's important to note that AMI IDs are different in every AWS region, so if you change the region parameter to something other than `us-east-2`, you'll need to manually look up the corresponding Ubuntu AMI ID for that region and copy it into the `ami` parameter.

- `instance_type`: The EC2 instance type determines the amount of CPU, memory, disk space, and networking capacity available. Each type offers different specifications, and the example provided uses `t2.micro`, which offers one virtual CPU and 1 GB of memory, and is included in the AWS Free Tier.

Step 2 – initialize Terraform

After saving the `main.tf` file in the newly created directory, the next step is to initialize Terraform. This process is similar to initializing a local repository using `git init`. The purpose of this step is to set up the Terraform environment and download any necessary plugins or modules. To initialize Terraform, open your terminal, navigate to the directory where the `main.tf` file is saved, and run the following command:

```
terraform init
```

The response should be similar to the following output:

```
Initializing the backend...
Initializing provider plugins...
- Checking for available provider plugins...
- Downloading plugin for provider "aws" (hashicorp/aws) 2.44.0...
The following providers do not have any version constraints in
configuration, so the latest version was installed.
To prevent automatic upgrades to new major versions that may contain
breaking changes, it is recommended to add version = "..." constraints
to the corresponding provider blocks in configuration, with the
constraint string suggested below.
* provider.aws: version = "~> 2.44"
Terraform has been successfully initialized!
You may now begin working with Terraform. Try running a "terraform
```

```
plan" to see any changes that are required for your infrastructure.
All Terraform commands should now work.

If you ever set or change modules or backend configuration for
Terraform rerun this command to reinitialize your working directory.
If you forget, other commands will detect it and remind you to do so
if necessary.
```

When working with Terraform, the code for providers such as AWS, Azure, and GCP are not included in the Terraform binary. Therefore, when starting with new Terraform code, it's essential to run the command `terraform init` to scan the code, identify which providers are being used, and download the corresponding code for them. By default, this code is downloaded into a `.terraform` folder, which serves as Terraform's scratch directory. Additionally, Terraform records information about the downloaded Provider code into a `.terraform.lock.hcl` file. It's important to note that the `init` command can be safely executed multiple times and is idempotent. In later chapters, we will explore further uses for the `init` command and `.terraform` folder.

Step 3 – pre-validate/predict the change—a pilot run

Running the `terraform plan -out tfplan` command will provide detailed information about the changes that will be made to your AWS infrastructure. The `-out tfplan` flag will save the `plan` output to a file named `tfplan`. This ensures that the planned changes will be applied without any modification, and what is seen during the planning phase will be committed. It is now time to apply the plan by running the `terraform apply tfplan` command:

```
terraform plan
```

The output of the previous command should display the changes that Terraform plans to make to your AWS infrastructure. It should look similar to the following:

```
(...)
Terraform will perform the following actions:
  # aws_instance.example will be created
  + resource "aws_instance" "example" {
      + ami                          = "ami-0fb653ca2d3203ac1"
      + arn                          = (known after apply)
      + associate_public_ip_address  = (known after apply)
      + availability_zone            = (known after apply)
      + cpu_core_count               = (known after apply)
      + cpu_threads_per_core         = (known after apply)
      + get_password_data            = false
      + host_id                      = (known after apply)
      + id                           = (known after apply)
      + instance_state               = (known after apply)
      + instance_type                = "t2.micro"
      + ipv6_address_count           = (known after apply)
```

```
        + ipv6_addresses                = (known after apply)
        + key_name                      = (known after apply)
        (...)
    }
Plan: 1 to add, 0 to change, 0 to destroy.
```

When you execute the `terraform plan` command, Terraform will provide a detailed output of what changes it plans to make to your AWS infrastructure. It's a great way to verify the resources that will be created or destroyed and to see whether anything unexpected will happen.

Keep in mind that when making modifications to existing resources, Terraform may need to destroy and recreate them. In such cases, the output will mention that the resource is going to be destroyed. Make sure to review the output carefully to avoid unintended results.

The `plan` command is an essential tool for verifying your Terraform code before applying it to your infrastructure. The output of the command is similar to the output of the `diff` command in Unix, Linux, and Git. The output displays a plus sign (+) for resources that will be created, a minus sign (-) for resources that will be deleted, and a tilde sign (~) for resources that will be modified in place.

In the case of the preceding output, Terraform is planning on creating a single EC2 instance and nothing else, which is exactly what you want. Be sure to monitor the last line of the output every time you run the `plan` command to ensure that there are no unintended results:

```
Plan: 1 to add, 0 to change, 0 to destroy.
```

Step 4 – apply the plan with terraform apply

Now that we have confirmed our changes with the `terraform plan` command, we can move forward and execute the changes using the `terraform apply` command. Unlike `terraform plan`, which is a dry run, `terraform apply` makes real changes to our AWS infrastructure based on the configuration file.

Be sure to double-check the output before entering yes to apply the changes. Once you have confirmed the changes, Terraform will begin creating the infrastructure resources:

```
terraform apply
```

The output should be as follows:

```
Terraform will perform the following actions:
  # aws_instance.example will be created
  + resource "aws_instance" "example" {
      + ami                          = "ami-0fb653ca2d3203ac1"
      + arn                          = (known after apply)
      + associate_public_ip_address  = (known after apply)
      + availability_zone            = (known after apply)
```

```
            + cpu_core_count              = (known after apply)
            + cpu_threads_per_core        = (known after apply)
            + get_password_data           = false
            + host_id                     = (known after apply)
            + id                          = (known after apply)
            + instance_state              = (known after apply)
            + instance_type               = "t2.micro"
            + ipv6_address_count          = (known after apply)
            + ipv6_addresses              = (known after apply)
            + key_name                    = (known after apply)
            (...)
      }
 Plan: 1 to add, 0 to change, 0 to destroy.
 Do you want to perform these actions?
    Terraform will perform the actions described above.
    Only 'yes' will be accepted to approve.

    Enter a value:
```

You'll observe that the `apply` command displays the same plan output and seeks your confirmation on whether you wish to proceed with said plan. Although the `plan` command is available separately, it's primarily useful for quick assessments and during code reviews. Most often, you'll directly execute the `apply` command and review the plan output it presents.

Type `yes` and hit *Enter* to deploy the EC2 instance:

```
 Do you want to perform these actions?
    Terraform will perform the actions described above.
    Only 'yes' will be accepted to approve.
    Enter a value: yes
 aws_instance.example: Creating...
 aws_instance.example: Still creating... [10s elapsed]
 aws_instance.example: Still creating... [20s elapsed]
 aws_instance.example: Still creating... [30s elapsed]
 aws_instance.example: Creation complete after 38s [id=i-07e2a3exxx]
 Apply complete! Resources: 1 added, 0 changed, 0 destroyed.
```

Congratulations! You have successfully deployed an EC2 instance in your AWS account using Terraform. To confirm this, go to the EC2 console and check that your instance has been provisioned.

Understanding AWS Providers

When working with Terraform to provision infrastructure in AWS, it's crucial to understand the concept of AWS Providers. In Terraform, a Provider is responsible for understanding the API interactions with a particular service and exposing the available resources and data sources. AWS is one of the most widely used cloud providers, and Terraform provides a rich set of AWS Provider resources to manage AWS infrastructure. In this section, we'll explore what AWS Providers are, how to configure and authenticate them, and best practices for working with AWS Providers in Terraform.

What are AWS Providers and why are they important in Terraform?

AWS Providers are plugins that allow Terraform to interact with the AWS API to manage infrastructure resources in AWS. They enable Terraform to provision, modify, and delete AWS resources such as EC2 instances, S3 buckets, and VPCs. Providers are critical components of Terraform, allowing it to automate the provisioning of infrastructure across multiple cloud platforms and on-premises data centers.

How to configure an AWS Provider in your Terraform code

Configuring an AWS Provider in Terraform is simple and straightforward. You just need to specify the AWS Provider and the region you want to work within your Terraform code. You can also optionally set your AWS access and secret keys as environment variables or use an AWS credentials file.

Here's an example of how to configure the AWS Provider in your Terraform code:

```
provider "aws" {
    region = "us-west-2"
}
```

In this example, we're using the aws Provider and setting the region to us-west-2. This means that any AWS resources we create with Terraform will be created in the US west (Oregon) region.

Once you have configured the AWS Provider in your Terraform code, you can start creating AWS resources using Terraform.

Understanding the different versions of the AWS Provider and their compatibility with Terraform

The AWS Provider for Terraform is versioned separately from Terraform itself. Each release of the AWS Provider includes new features, bug fixes, and compatibility updates for new AWS services and features. It's important to check the compatibility of the AWS Provider with your version of Terraform before upgrading.

When using a version of Terraform that is not compatible with the AWS Provider, you may experience issues such as errors when running Terraform commands or unexpected behavior when deploying resources.

To check the compatibility of the AWS Provider with your version of Terraform, you can refer to the AWS Provider release notes or the Terraform documentation. In general, it's recommended to always use the latest version of the AWS Provider that is compatible with your version of Terraform to take advantage of the latest features and bug fixes.

Best practices for working with AWS Providers in Terraform

Here are some best practices for working with AWS Providers in Terraform:

1. Keep your AWS Provider version up to date to ensure compatibility with the latest features and bug fixes.
2. Use separate profiles for each Terraform workspace in order to have different AWS credentials for different environments.
3. Use AWS IAM roles and policies to restrict access to your resources and use the least-privilege principle.
4. Use Terraform's `plan` and `apply` commands to test changes before deploying them to production.
5. Use modules to encapsulate and reuse Terraform code, including AWS Provider configurations.
6. Follow the principle of least configuration, and avoid configuring unnecessary settings in your AWS Provider block.
7. Use Terraform Cloud or Terraform Enterprise to securely store and manage your AWS credentials, as well as to collaborate with your team on infrastructure changes.

Understanding Terraform modules

Terraform modules are reusable, encapsulated packages of Terraform code that allow you to efficiently manage and organize your infrastructure. They help you to abstract common infrastructure patterns, reduce code duplication, and make it easier to maintain, update, and share your infrastructure code. In this section, we'll dive into the details of Terraform modules and learn how to use them effectively to manage your infrastructure.

What is a Terraform module?

Terraform modules are powerful features that allow you to encapsulate a group of resources dedicated to one task into a collection of standard configuration files within a dedicated directory. This reduces the amount of code needed for similar infrastructure components and makes it easier to manage and reuse configuration code. When you run Terraform commands from a module directory, it is considered the root module. In fact, every Terraform configuration is part of a module. The following is an example of a simple set of Terraform configuration files:

```
.
├── LICENSE
├── README.md
├── main.tf
├── variables.tf
├── outputs.tf
```

Figure 5.4 – Terraform configuration files

If you are running Terraform commands from within the `minimal-module` directory, the contents of that directory are considered the root module. This means that the files in this directory define a single module, which could contain one or more resources.

Using modules

When working with Terraform, it's important to understand how to organize your code to manage complexity and reuse code across different projects. One way to achieve this is by using Terraform modules. A module is essentially a collection of configuration files in a dedicated directory that encapsulates groups of resources dedicated to one task, reducing the amount of code you must develop for similar infrastructure components. These modules can be called from other directories through module blocks, allowing you to reuse code across different projects. In this context, a module that is called by another configuration is referred to as a child module.

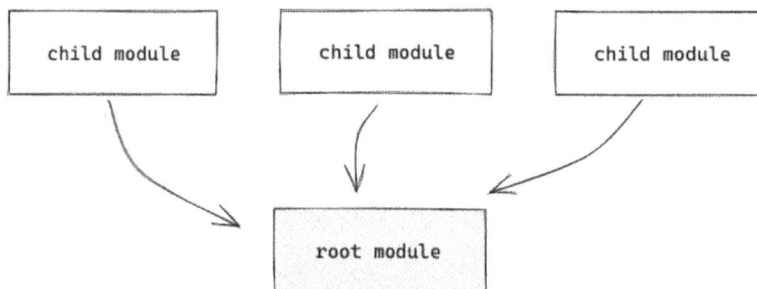

Figure 5.5 – Child module

Local and remote modules

Terraform modules can be loaded from either a local file system or a remote source. Remote sources supported by Terraform include the Terraform Registry, several version control systems, HTTP URLs, and private module registries in Terraform Cloud or Terraform Enterprise.

Module best practices

Using Terraform modules is essential for creating reusable and maintainable infrastructure code. They provide a way to encapsulate related resources into a single component and can be used to share common patterns and best practices across teams and projects. To get the most out of modules, it is recommended to follow these best practices:

1. It is important to follow a naming convention while naming your Terraform Provider `terraform-<PROVIDER>-<NAME>`. This convention must be followed if you plan to publish your Provider to the Terraform Cloud or Terraform Enterprise module registries.

2. Consider using modules when designing and writing your Terraform configuration, even for smaller projects. Even if you are the only person working on the configuration, the benefits of using modules can save time and effort in the long run.

3. To organize your code and reduce the burden of maintaining and updating your configuration as your infrastructure grows in complexity, it is recommended to use local modules. This is beneficial even if you are not using or publishing remote modules. Therefore, it is best to organize your configuration with modules from the beginning.

4. Leverage the public Terraform Registry to discover useful modules. This will help you implement your configuration more efficiently and confidently, as you can rely on the pre-built modules to implement common infrastructure scenarios instead of building everything from scratch.

5. Collaboration is a key aspect of infrastructure management, and modules enable teams to work together effectively to create and maintain infrastructure. To enhance collaboration, you can publish and share modules with your team. You can publish modules publicly on the Terraform Registry or privately through Terraform Cloud or Terraform Enterprise. Module users can then reference the published child modules in their root module or deploy no-code ready modules through the Terraform Cloud UI.

What problems do Terraform modules solve?

When working with Terraform, managing large and complex infrastructure can be a daunting task. Terraform modules provide a solution to this problem by encapsulating groups of resources and configurations into reusable and shareable components. In this section, we'll explore the various problems that Terraform modules solve and how they can benefit your infrastructure management workflow:

*   **Code repetition**: As your Terraform infrastructure grows larger, copying and pasting code becomes inefficient and time-consuming. When you need to create multiple instances of the same resource, repeating the code is not scalable. It leads to code repetition, which is not only time-consuming but also increases the chances of human error. Terraform modules solve this problem by encapsulating groups of resources dedicated to one task and reducing the amount of repetitive code you need to write.

- **Lack of code clarity**: Copying and pasting code is not only inefficient; it also makes the code base difficult to maintain and understand. When working with large-scale infrastructure in Terraform, a modular approach can help address this issue. Using modules dedicated to specific tasks allows for a more organized and readable code base that is easier to maintain and understand.

- **Lack of compliance**: Creating a Terraform module in accordance with best practices ensures that the same pattern is followed whenever it is reused. Whether it's for encryption, redundancy, or lifecycle policies, practices configured inside the module will be enforced, eliminating the need to repeat the process manually.

- **Human error**: Creating a group of resources from scratch or copy-pasting them can lead to errors such as renaming or overwriting something. Terraform modules provide a solution to this problem by allowing you to create a single module, test it, and reuse it in multiple places. This approach ensures that all elements are correct and consistent throughout your infrastructure. By using a single block, it becomes easier to check and test your code. Terraform modules also provide other benefits, but it's important not to overuse them. It's essential to find the right balance and maintain it.

How to implement best practices with Terraform AWS modules

When working with AWS infrastructure, it's important to follow best practices to ensure a reliable, scalable, and secure environment. Terraform AWS modules provide a way to implement these best practices efficiently and consistently across different environments. In this section, we will explore some of the best practices for using Terraform AWS modules and how to implement them in your infrastructure. We'll cover topics such as module organization, naming conventions, versioning, and more.

Terraform configurations file separation

Storing all Terraform code in a single file such as `main.tf` can make it challenging to read and maintain the code. A better approach is to split the code across multiple files, each dedicated to a specific purpose or resource. This not only makes the code more organized but also easier to troubleshoot and update in the future:

- `main.tf`: This file calls modules, locals, and data sources to create all the necessary resources

- `variables.tf`: Declarations of variables used in `main.tf` are included in this file

- `outputs.tf`: This file contains outputs from the resources created in `main.tf`

- `versions.tf`: This file specifies version requirements for Terraform and providers

- `terraform.tfvars`: This file contains variable values and should not be used anywhere else

Follow a standard module structure

The standard module structure must be followed for Terraform modules:

- It is recommended to group resources based on their shared purpose in separate files such as `vpc.tf`, `instances.tf`, or `s3.tf` rather than creating individual files for each resource

- Ensure that every module includes a Markdown-formatted `README.md` file containing essential documentation about the module

Use opinionated modules to do exactly what you need

When creating Terraform modules, it's recommended to make them opinionated for your specific use case, unless you intend to publish them as open source or for general-purpose use. You can make use of existing resources, open source modules, or even create your own. However, be cautious about creating too many module dependencies, as it can become difficult to maintain and update your code.

Leverage official open source modules

Consider using open source modules provided freely by HashiCorp and the platform you are using. These modules can be used as primitives by modules you create, or they can be used in your deployments as they come if they achieve everything you need. You just need to ensure you call specific versions of them so your deployments are consistent.

I've seen many people who advocate forking open source modules and tweaking them. I'd be cautious when following that approach for three reasons:

- Forking an open source repository and changing it means you are now the maintainer of that module, giving yourself and other members of your team a higher workload

- Engineers are likely to be familiar with an open source module, but not your bespoke version of it, hence new staff enrolment would be quicker if you are using standard modules

- Open source modules are generally too broad; your in-house modules should be opinionated for your use case to make them simpler to use and maintain

Forking is sometimes mandatory; some companies such as banks require you to fork modules and keep them in-house, but if you do that, you should consider not changing them at all and just tracking the official version and updating where possible.

That all being said, there may be situations when importing an official module and changing it may suit your needs. In that case, I recommend you strip it bare of all the features you don't need and simplify it as much as possible.

Also, in that case, consider using open source modules that are designed to be forked if you can find them. For example, Google Cloud has the cloud foundation fabric modules that are designed just for this purpose.

Make extensive use of convention over configuration

Your modules will be opinionated to do what you need to do; hence you should default as many variables as possible and only require the bare minimum for setup. This will help keep your deployment code clean and easy to understand and change. Ideally, you should only need five or six variables at most. Default everything else if you can.

Make modules flexible with multiple optional inputs

You should be able to use your modules with minimal inputs, but that doesn't mean that they shouldn't be flexible for changes. This minimizes the need to have to change code, and it gives you options based on different situations. That being said, don't get too bogged down trying to predict every scenario under the sun; start by making things as simple as possible based on your use case.

Refer to modules by version

Don't avoid specifying the version of your modules, as that could break your deployments whenever you make changes to them. Consider using semantic versioning to update your modules.

Consider bundling modules together if they serve a common purpose

At the start of a project, you may consider keeping your modules and deployments in the same repository and refer to them by path. As your product matures, you may want to move these modules to a separate repository to be able to refer to them by version and maintain them separately.

Having one repository per module is useful if your modules need to be maintained by different teams or if they are common modules used by multiple projects. However, if you have a set of closely related modules, consider keeping them all in one repository and using them as submodules.

You can still version them in this way and it is much easier to manage them. Remember that you can always separate the modules later in their own repositories, but always start with the simplest setup possible.

Consider using variable and naming validation

Terraform has a relatively new feature where you can validate names with **Regular Expressions** (**regex**). This is very useful to avoid naming errors with your platform before you hit **Apply**. You could also enforce the way your resources are named by concatenating inputs such as labels and prefixes while validating each of them, keeping your platform naming consistent.

```
### variables.tf ####

variable "csp" {
  type        = string
  description = "Three letter word denoting cloud service provider. ie. gcp "
  validation {
    condition     = can(regex("^(gcp|aws|azu)$", var.csp)) == true
    error_message = "The csp (cloud service provider) variable value can only be gcp, aws or azu."
  }
  default = "gcp"
}
```

Figure 5.6 – Naming validation

Use locals correctly

I've seen locals in situations where variables would be better suited. I find locals to be very useful, especially in the following situations:

- Using functions on your outputs and/or inputs
- Concatenating variables to form names of resources
- Using conditional expressions

You can keep locals in their own file, but I generally recommend keeping them in the same file, close to the code they are used for.

Keep the code in your module logically separated

I generally advocate keeping the structure of the files standard to avoid confusion. However, if your module requires over 200 lines of code, not including variables, you should consider splitting the main.tf into multiple files according to what they do and keep all related resources and locals within that file. This makes it easier to modify and read than having to search through a long main.tf file—even if that file is separated with comment lines.

Separate required and optional variables

To improve the readability of your code, keep required variables at the top and optional variables at the bottom. Separate them with a comment line in your variables.tf file.

Always have an example folder within your module folder

An example folder has two advantages:

- It gives users an idea of how to use your module in a deployment

- You can use it to test your module code before creating a new version for it

Summary

You have successfully learned about Terraform modules and AWS Providers. These are essential tools that will help you manage and deploy your infrastructure in AWS using Terraform. Now that you understand how to use modules and providers, the next step is to make decisions for your projects in AWS Terraform.

In the next chapter, you will learn how to make decisions about the architecture, security, and scalability of your projects. You'll also explore how to use best practices for cost optimization and how to manage your infrastructure as code. With these skills, you'll be able to design and implement robust, efficient, and cost-effective infrastructure for your projects in AWS using Terraform. So, get ready to take your Terraform skills to the next level and create the best infrastructure for your projects in AWS.

6
Making Decisions for Terraform Projects with AWS

Welcome to the next chapter of our journey with AWS Terraform. By now, you've learned about the importance of Terraform modules and AWS providers and how to implement best practices for your Terraform code. In this chapter, we'll be diving deeper into the fundamentals of the **Amazon Web Services** (**AWS**) infrastructure, network, and resources. With these essential skills, you'll be better equipped to make informed decisions about your AWS Terraform projects.

We'll also explore the decision-making process when using templates or modules and how to structure your projects using AWS environments, projects, and components. By the end of this chapter, you'll have the knowledge and skills necessary to make sound infrastructure decisions for your AWS Terraform projects. So, let's dive in!

Here are the main topics we will cover:

- AWS infrastructure and fundamentals
- AWS Organizations and network fundamentals
- AWS resources fundamentals
- AWS environments, projects, workloads

AWS infrastructure and fundamentals

When working with Terraform to manage your infrastructure on AWS, it is essential to understand the basics of AWS infrastructure. In this section, we will cover the fundamentals of AWS infrastructure and how they relate to Terraform. We will explore the building blocks of AWS infrastructure, including AWS Regions, Availability Zones, and VPCs, and discuss how to design and plan your infrastructure using Terraform. With a strong understanding of AWS infrastructure, you will be better equipped to make informed decisions when creating and managing your infrastructure with Terraform.

What is AWS infrastructure?

AWS is a comprehensive cloud computing platform that provides a wide range of services, such as storage, networking, analytics, machine learning, and much more. It is a highly scalable and reliable platform that caters to the needs of businesses of all sizes. The infrastructure fundamentals of AWS refer to the basic building blocks and underlying technology that support these services. These include servers, storage, networking, and data centers. AWS allows businesses and organizations to access scalable, reliable, and secure infrastructure services on a pay-as-you-go basis without the need for them to invest in and maintain their own physical infrastructures. This can help reduce costs and increase flexibility and agility.

What is infrastructure as a service?

In the **infrastructure-as-a-service (IaaS)** cloud computing model, virtualized computing resources are provided over the internet. Providers such as AWS offer organizations access to infrastructure services such as servers, storage, networking, and data centers on a pay-as-you-go basis. This eliminates the need for businesses to invest in and maintain their own physical infrastructure, allowing them to access scalable, reliable, and secure infrastructure on demand.

What is platform as a service?

In the cloud computing model of **platform as a service (PaaS)**, providers such as AWS offer businesses and organizations a complete platform to develop, test, deploy, and manage software applications. The PaaS platform includes operating systems, middleware, databases, and other services, freeing businesses from the burden of managing and maintaining the underlying infrastructure. This enables organizations to focus on building and improving their applications while relying on the PaaS provider to manage the platform.

What is software as a service?

In the **software as a service (SaaS)** cloud computing model, users can access software applications hosted and managed by a provider through the internet without needing to install or maintain the application themselves. This allows businesses and organizations to focus on using the software instead of managing it, and SaaS providers typically charge users on a subscription basis. This model has become increasingly popular due to its scalability, flexibility, and cost-effectiveness.

AWS offers a range of IaaS, PaaS, and SaaS products and services. The following are some of them:

Some AWS IaaS services

- **Amazon Elastic Compute Cloud (EC2)**: Offers flexible and scalable virtual servers in the cloud
- **Amazon Elastic Container Service (ECS)**: Enables Docker container management and orchestration on EC2 instances

- **Amazon Elastic Container Service for Kubernetes**: Facilitates the deployment and management of Kubernetes clusters on AWS

- **Amazon Elastic Kubernetes Service (EKS)**: Allows businesses to create and operate Kubernetes clusters on AWS

- **Amazon LightSail**: Provides simple **virtual private servers (VPS)** for web development and small-scale applications

- **Amazon Elastic Block Store (EBS)**: Provides persistent block storage volumes for EC2 instances

- **Amazon Elastic File System (EFS)**: Offers scalable and shared file storage for use with EC2 instances

- **Amazon S3**: Provides highly scalable and durable object storage for data and files

- **Amazon Glacier**: Provides low-cost archival storage for data retention and retrieval

- **Amazon CloudFront**: Delivers content globally through a fast and secure **content delivery network (CDN)**

- **Amazon Route 53**: Offers a scalable and reliable **domain name system (DNS)** service for managing DNS records

- **Amazon Virtual Private Cloud (VPC)**: Allows businesses to create their own isolated virtual network in the AWS cloud

- **AWS Direct Connect**: Allows businesses to establish a dedicated network connection between their on-premises infrastructure and AWS

Some AWS PaaS services

- **AWS Elastic Beanstalk**: Simplifies the deployment and management of web applications on AWS

- **AWS Lambda**: Enables developers to run code without managing servers or infrastructure

- **AWS CodePipeline**: Automates the build, test, and deployment of code changes

- **AWS CodeBuild**: Offers a fully managed build service for compiling source code into deployable artifacts

- **AWS CodeDeploy**: Automates application deployments to compute instances, on-premises servers, or AWS Lambda

- **AWS CodeStar**: Provides a unified interface for managing the entire application development lifecycle on AWS

- **AWS CloudFormation**: Allows organizations to define and manage AWS resources as code using templates

- **AWS CloudTrail**: Records API activity and delivers log files for auditing and compliance purposes
- **AWS X-Ray**: Facilitates the tracing, debugging, and analysis of distributed applications running on AWS.

Some AWS SaaS services

- **Amazon WorkSpaces**: Offers cloud-based virtual desktops that can be accessed by remote and mobile workers
- **Amazon Chime**: Provides a cloud-based platform for communication and collaboration through messaging, meetings, and video conferencing
- **Amazon Connect**: Offers a cloud-based contact center platform for businesses
- **Amazon AppStream 2.0**: Enables businesses to stream desktop applications to users over the internet
- **Amazon WorkDocs**: Provides a cloud-based content management and collaboration platform for businesses
- **Amazon WorkMail**: Offers a cloud-based email and calendar service for businesses
- **Amazon Elasticsearch Service**: Enables businesses to deploy, operate, and scale Elasticsearch clusters easily in the cloud
- **Amazon Kendra**: Provides a machine learning-powered enterprise search service
- **Amazon Managed Blockchain**: Enables businesses to easily create and manage scalable blockchain networks
- **Amazon Quantum Ledger Database (QLDB)**: Provides a fully managed ledger database for applications that need a central, trusted authority to maintain a complete and verifiable record of transactions

What are the main AWS product and service categories?

AWS provides a variety of cloud computing products and services organized into different resource categories. These include the following:

- **Compute**: Provides services for running and managing compute resources, such as virtual machines and containers. Examples include EC2, Amazon ECS, and AWS Lambda.
- **Storage**: Provides services for storing and managing data, such as files and objects. Examples include Amazon S3, EBS, and Amazon EFS.
- **Database**: Provides services for running and managing databases in the cloud. Examples include Amazon Aurora, Amazon DynamoDB, and Amazon Redshift.

- **Networking**: Provides services for networking, connectivity, and content delivery. Examples include Amazon VPC, Amazon Route 53, and Amazon CloudFront.

- **Security and Identity**: Provides services for securing and managing access to AWS resources. Examples include AWS **Identity and Access Management (IAM)**, AWS **Key Management Service (KMS)**, and Amazon GuardDuty.

- **Analytics**: Provides services for collecting, processing, and analyzing data. Examples include Amazon EMR, Amazon Kinesis, and Amazon Athena.

- **Machine Learning**: Provides services for building and deploying machine learning models. Examples include Amazon SageMaker, Amazon Rekognition, and Amazon Lex.

- **Management Tools**: Provides services for managing and optimizing AWS resources. Examples include AWS CloudFormation, AWS CloudWatch, and AWS Trusted Advisor.

These and other resource categories provided by AWS can help businesses and organizations access a wide range of cloud computing services to support their operations and goals.

How to make decisions to start a Terraform project in AWS

When deciding to use Terraform for an AWS project, there are a few key considerations to keep in mind. These include the following:

- **The scope and scale of the project**: Terraform is designed for IaaC and can be used to manage infrastructure for large, complex projects. If your project involves multiple AWS services and resources, Terraform can help you manage them efficiently and reliably.

- **The level of automation and integration required**: Terraform allows you to automate the provisioning and management of your AWS infrastructure using configuration files and declarative syntax. This can help reduce manual errors and improve consistency across your environment. Terraform also integrates with other AWS services and tools, such as AWS CloudFormation, AWS CodePipeline, and AWS CodeBuild.

- **The level of collaboration and team size**: Terraform supports collaborative infrastructure management through the use of version control systems such as Git. This can help teams work together more efficiently and effectively, and it also enables you to track and roll back changes to your infrastructure. If you have a large team working on your project, Terraform can help you manage and coordinate their efforts.

- **The level of support and documentation available**: Terraform is an open source tool with a large and active community. This means there is a wealth of documentation, tutorials, and other resources available to help you learn and use Terraform effectively. AWS also provides its own documentation and support for Terraform, including best practices and integration with other AWS services.

How to start designing your first AWS infrastructure

There are several key steps involved in designing AWS infrastructure, which are as follows:

1. Identify and define your business requirements and goals for the organization's infrastructure. This will help you understand what your infrastructure needs to do and how it needs to support your business operations for your organization.

2. Decide on the number and type of AWS accounts you need. This will likely depend on factors such as the size and complexity of your business, the number of teams and users who need access to AWS, and the level of security and compliance requirement.

3. Select the appropriate AWS services and resources for your infrastructure. This will likely involve choosing from a range of compute, storage, and networking services based on your requirements and goals.

4. Plan and design your infrastructure architecture. This will involve creating diagrams and other visualizations that show how the different AWS services and resources will be connected and configured to support your business needs.

5. Implement and deploy your infrastructure. This will involve using tools and services such as AWS CloudFormation, AWS CodePipeline, and AWS CodeBuild to automate the provisioning and configuration of your AWS resources.

6. Monitor and maintain your infrastructure. This will involve using tools such as AWS CloudWatch and AWS Trusted Advisor to monitor the performance and health of your infrastructure to address any issues or potential improvements and using AWS Organizations, AWS **Single Sign-On** (**SSO**), and AWS IAM to manage and monitor the security, compliance, and usage of your AWS accounts.

By following these steps, you can design a well-planned, scalable, and reliable AWS infrastructure that meets the needs of your business.

AWS Organizations and network fundamentals

- **AWS Organizations** is a service that allows businesses and organizations to manage and govern their AWS accounts in a centralized and scalable manner. AWS Organizations enables you to create and manage a hierarchy of AWS accounts and apply policies across your accounts to help ensure compliance with corporate standards and best practices. This can help you manage your AWS infrastructure and resources more efficiently, reduce the risk of errors and security vulnerabilities, and improve the visibility and control of your AWS usage. AWS Organizations is available through the AWS Management Console, the AWS CLI, or the AWS Organizations API.

- An **AWS account** is a user-defined entity that provides access to the services and resources provided by AWS. An AWS account is the starting point for using AWS and is used to identify and authenticate users who want to access AWS services and resources. AWS accounts are created and managed through the AWS Management Console, which is the web-based interface for accessing and managing AWS services. AWS accounts are typically associated with an email address and a password and can be accessed using the AWS Management Console or the AWS **Command Line Interface (CLI)**. AWS accounts can be used individually or as part of an AWS Organizations structure to manage and govern multiple AWS accounts in a centralized and scalable manner.

- An **AWS Region** is a physical location around the world where AWS provides services and resources. AWS Regions are composed of multiple Availability Zones, which are isolated, fault-tolerant data centers that provide low-latency connectivity to end users. AWS Regions are designed to be redundant and highly available and are used to host and run the various services and resources provided by AWS. AWS customers can choose the AWS Region that best meets their performance, compliance, and other requirements and can access and use the services and resources in that region through the AWS Management Console, the AWS Command-Line Interface (**CLI**), or the AWS **application programming interface** (**API**).

- AWS **Availability Zones** are isolated, fault-tolerant data centers that provide low-latency connectivity to end users. AWS Availability Zones are located within AWS Regions, which are physical locations around the world where AWS provides services and resources. Each Availability Zone is composed of one or more data centers and is designed to be redundant and highly available. This means that if one data center in an Availability Zone goes down, the other data centers in the same Availability Zone will continue to operate, so users will not experience any interruption in service. AWS customers can use Availability Zones to run their applications and workloads with high availability and fault tolerance and can choose the Availability Zones that best meet their performance and compliance requirements.

- **Amazon VPC** is a cloud computing service provided by AWS that allows businesses and organizations to create and configure their own virtual private network in the AWS cloud. A VPC enables you to define and customize your own network settings, including the IP address range, subnets, route tables, and network gateways. This allows you to create a logically isolated and secure network environment that is separate from the rest of the AWS cloud. VPCs can be used to host and run AWS services and resources, such as Amazon EC2 instances, Amazon EBS volumes, and Amazon S3 buckets. VPCs can also be connected to your on-premises infrastructure using AWS Direct Connect or a VPN connection, allowing you to seamlessly extend your own network into the AWS cloud.

- An **AWS subnet** is a range of IP addresses within an Amazon VPC that is associated with a specific Availability Zone. Subnets are used to organize and segment the network within a VPC and can be used to control the traffic between different groups of AWS resources. Each subnet is associated with a route table, which specifies the traffic flows within and between the subnet and other network destinations. Subnets can be public or private, depending on whether they have internet connectivity. Public subnets are connected to the internet through an internet gateway, while private subnets are not directly connected to the internet and can only access the internet through a NAT gateway or VPN connection. AWS customers can use subnets to design and implement a scalable and secure network architecture within their VPCs.

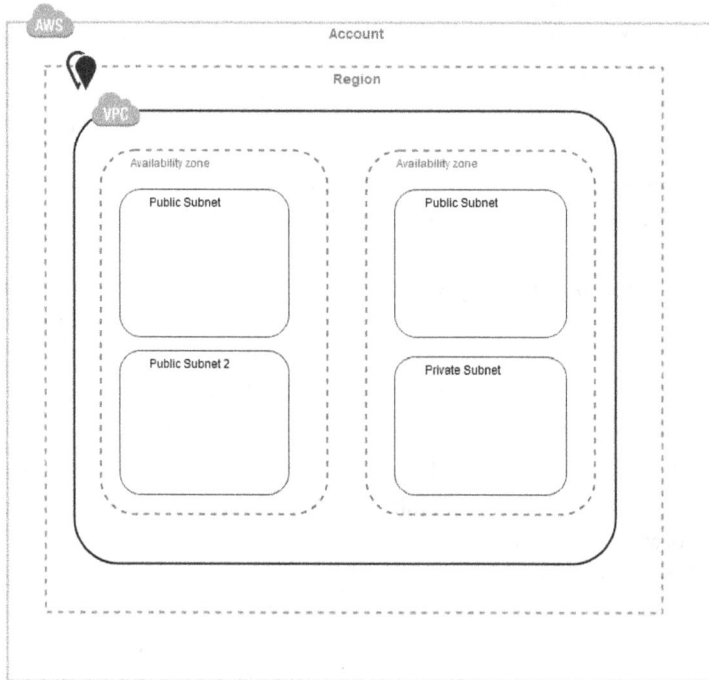

Figure 6.1 – AWS subnet

AWS resources fundamentals

The AWS core principles refer to the fundamental values and beliefs that guide the design and operation of AWS. These principles include the following:

- **Customer obsession**: AWS is focused on meeting the needs and exceeding the expectations of its customers

- **Innovation**: AWS is committed to continuous innovation and improvement in its products and services

- **Global infrastructure**: AWS operates a global network of data centers and regions to provide customers with low-latency and high-availability services

- **Responsiveness**: AWS is designed to be fast and responsive, allowing customers to quickly and easily access and use its services

- **Operational excellence**: AWS is focused on delivering high-quality, reliable, and secure services to its customers

- **Security**: AWS is committed to the security and privacy of its customers' data and resources

- **Cost-effectiveness**: AWS is designed to provide customers with a cost-effective and flexible way to access and use its services

By following these principles, AWS can deliver a wide range of cloud computing services that meet the needs of its customers and enable them to achieve their goals. While starting a new project, you should consider the same principles so you can deliver the best solution for your infrastructure.

AWS shared responsibility model

The AWS shared responsibility model is a framework that defines the roles and responsibilities of AWS and its customers with respect to the security and compliance of their AWS resources. Under the AWS shared responsibility model, AWS is responsible for the security of the cloud infrastructure, including the physical security of its data centers, the security of its network and hardware, and the security of its services and features. AWS customers, on the other hand, are responsible for the security of their own applications and data, as well as the configuration of their AWS resources. This means that customers are responsible for protecting their own data and applications from threats and vulnerabilities and for ensuring that their AWS resources are properly configured and used in compliance with their own security and compliance policies. The AWS shared responsibility model helps customers understand their own security and compliance responsibilities and enables them to design and implement a secure and compliant AWS environment that meets their business needs.

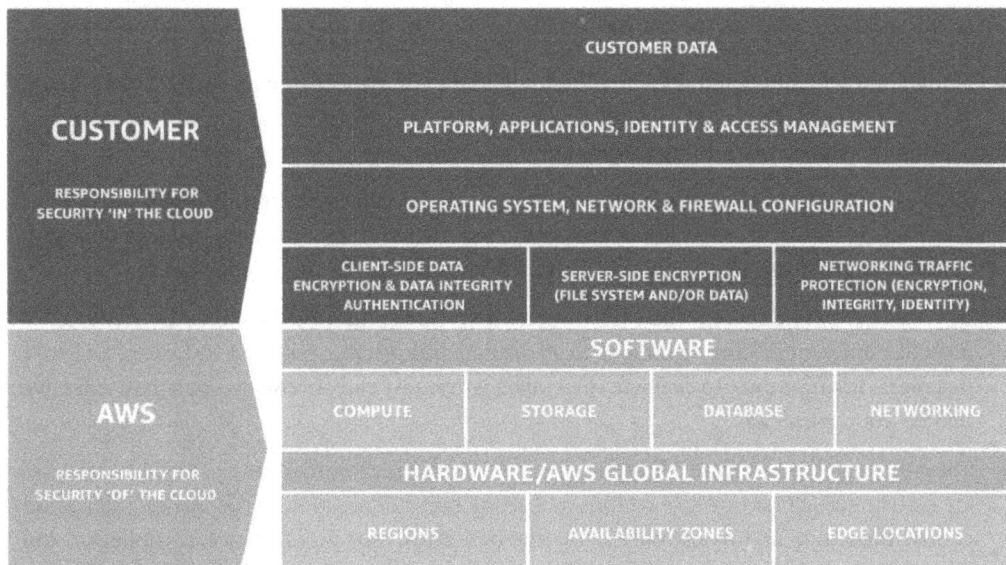

Figure 6.2 – AWS shared responsibility model

The shared responsibility model applies to all three types of cloud computing service models: IaaS, PaaS, and SaaS. Under the shared responsibility model, the cloud provider (such as AWS) is responsible for the security of the underlying infrastructure and infrastructure components, while the customer is responsible for the security of their own applications, data, and other resources that are built on top of the infrastructure.

In the case of IaaS, the cloud provider is responsible for the security of the physical infrastructure (e.g., servers, storage, and networking), the virtualization infrastructure (e.g., hypervisors and virtual machines), and the infrastructure services (e.g., identity and access management, and network security). The customer is responsible for the security of their own applications, data, and operating systems that run on the infrastructure.

In the case of PaaS, the cloud provider is responsible for the security of the underlying infrastructure, as well as the platform components and services that are provided as part of the PaaS offering (e.g., databases, application runtime environments, and load balancers). The customer is responsible for the security of their own applications and data that are built on top of the platform.

In the case of SaaS, the cloud provider is responsible for the security of the underlying infrastructure, the platform, and the SaaS application itself. The customer is responsible for the security of their own data and user access to the SaaS application.

In all cases, the shared responsibility model means that the cloud provider and the customer both have a role to play in ensuring the security and compliance of their AWS resources.

How to select AWS resources

When selecting AWS resources, there are several key factors to consider, including the following:

- Consider the business requirements and goals of your project or workload. This will help you determine the type and quantity of resources you need, as well as the performance and availability requirements for those resources.

- AWS offers a diverse array of services and features, each with its own strengths and limitations, that you can choose from to suit your specific requirements. It's crucial to conduct thorough research and compare the various options to determine the best fit for your needs.

- It's essential to understand the various pricing options and models offered by AWS. AWS provides different pricing options such as on-demand pricing, reserved instances, and spot instances. It's important to evaluate these pricing options and choose the most cost-effective one for your workload.

- The AWS regional and Availability Zone availability of the resources you need. AWS resources are distributed across multiple regions and Availability Zones around the world. You should choose the region and Availability Zone that best meet your performance, compliance, and other requirements.

- There may be AWS resource limits and quotas that apply to the resources you want to use. AWS imposes certain limits and quotas on the use of its services and resources to ensure the performance and availability of the AWS cloud for all customers. You should check the limits and quotas that apply to the resources you want to use and plan accordingly.

By considering these and other factors, you can select the appropriate AWS resources for your project or workload and use them effectively to support your business goals.

AWS environments, projects, workloads

In order to effectively manage and organize infrastructure on AWS using Terraform, it's important to understand the concept of environments, projects, and workloads. In this chapter, we will dive into the details of these concepts and explore best practices for implementing them in your Terraform code.

What is an environment?

Environments refer to the infrastructure and tools used to support the software development and deployment process in a DevOps model. DevOps is an approach to software development that prioritizes collaboration, automation, and continuous improvement in the development, testing, and deployment of software. DevOps environments are typically composed of a combination of on-premises and cloud-based resources and may include tools and services such as version control systems, **continuous integration and delivery (CI/CD)** tools, testing frameworks, and infrastructure automation tools. DevOps environments are designed to enable teams to rapidly and reliably develop, test, and deploy software, and support the continuous iteration and improvement of their software. By using DevOps environments, teams can improve the speed, quality, and reliability of their software development and deployment processes.

How to define environments or projects in AWS

There are several ways to define environments or projects in AWS depending on your specific requirements and goals. Some common approaches include the following:

- **Using AWS accounts**: You can create separate AWS accounts for each environment or project and use AWS Organizations to manage and govern the accounts. This approach provides a high degree of isolation and security between the different environments or projects and allows you to apply different policies and controls to each account.

- **Using VPCs**: You can create separate Amazon VPCs for each environment or project and configure the VPCs with different settings and resources (e.g., subnets, security groups, and route tables). This approach allows you to isolate the network and security settings for each environment or project and use different AWS resources in each VPC.

- **Using resource names and tags**: You can use unique names and tags for each resource in each environment or project and use AWS resource policies and IAM policies to control access to the resources based on their names and tags. This approach allows you to manage and access the resources in each environment or project independently and apply different access controls to each environment or project.

- **Using deployment environments**: You can use AWS CodePipeline to create and manage separate deployment environments for each environment or project (e.g., development, staging, and production). This approach allows you to automate the deployment and release of your application code and resources and control the promotion of code and resources between environments.

By using one or more of these approaches, you can effectively define and manage your environments or projects in AWS.

Summary

In this chapter, we learned about the fundamentals of AWS infrastructure, network, and resources, as well as how to decide when to use templates or modules. We also explored how to organize projects into environments, projects, and workloads. By the end of the chapter, we gained skills in AWS infrastructure decisions, AWS network fundamentals, and AWS resource fundamentals, and we learned how and when to develop modules and templates.

In the next chapter, we will dive into implementing Terraform in projects, where we will explore how to structure Terraform code, manage state, and use remote backends. Stay tuned!

7

Implementing Terraform in Projects

Are you ready to start developing your AWS infrastructure with Terraform? In this chapter, you'll learn the basics of Terraform and how to deploy your first template in AWS. We'll cover the process of selecting the right AWS provider and choosing public modules that meet your project's needs. You'll also learn how to write custom Terraform AWS modules for your specific use cases.

By the end of this chapter, you will have developed and deployed your AWS infrastructure using Terraform. You'll also have gained valuable skills in decision-making for providers and selecting the right public modules for your project's needs. Additionally, you'll learn how and when to develop custom AWS modules and how to use them effectively.

In this chapter, we will delve deeper into Terraform and explore how it can be used to develop and deploy AWS infrastructure projects. Here's a look at the topics we'll cover:

- Terraform basics for developing AWS infrastructure projects
- Selecting AWS Providers
- Selecting AWS public modules for your needs
- How to write custom Terraform AWS modules

Terraform basics for developing AWS infrastructure projects

Terraform is a tool for building, changing, and versioning infrastructure safely and efficiently. It can manage infrastructure for many different types of cloud providers, including AWS.

Let's look at some basic concepts in Terraform.

Resources

A resource is an element of your infrastructure, such as an EC2 instance, an S3 bucket, or a security group.

A resource is typically created using a resource block in your Terraform configuration. A resource block has a type and a name, and it specifies the desired state of the resource. For example, the following block creates an EC2 instance in AWS:

```
resource "aws_instance" "web_server" {
    ami           = "ami-12345678"
    instance_type = "t2.micro"
}
```

This resource block creates an EC2 instance with the type `aws_instance` and the name `web_server`. It specifies that the instance should be created using the specified AMI and instance type.

When you run `terraform apply`, Terraform will create the EC2 instance and set it to the desired state specified in the resource block. If the EC2 instance already exists and its state differs from the desired state, Terraform will update the instance to match the desired state.

You can also use resource properties to specify additional details about the resource, such as the VPC in which it should be created, the security groups it should be associated with, and so on.

Providers

A provider is a plugin that Terraform uses to interact with the infrastructure of a particular cloud provider, such as AWS. Each provider has its own set of resources that you can use to create and manage infrastructure.

To use a provider in your Terraform configuration, you need to specify it in a provider block. For example, to use the AWS provider, you would add the following block to your configuration:

```
provider "aws" {
    region = "us-east-1"
}
```

This block specifies that you want to use the AWS provider and that you want to use the us-east-1 region. You can also specify provider-specific configuration options, such as the access key and secret key to use when authenticating with the provider's APIs.

Once you've specified a provider in your configuration, you can use resources from that provider to create and manage infrastructure. For example, you could use the `aws_instance` resource to create an EC2 instance in AWS:

```
resource "aws_instance" "web_server" {
  ami           = "ami-12345678"
  instance_type = "t2.micro"
}
```

State

Terraform maintains a state file that stores the current configuration of your infrastructure. This allows it to track changes and know what actions to take when you make changes to your configuration.

The state file is an important part of how Terraform functions, as it allows Terraform to know what actions to take when you make changes to your infrastructure. For example, if you create a new EC2 instance using Terraform, the state file will be updated to reflect the existence of the new instance. If you then modify the instance using Terraform, the state file will be updated to reflect the new configuration of the instance.

There are a few different ways that you can store your state file:

- **Local state file**: By default, Terraform will store the state file locally in a file named `terraform.tfstate`. This is the simplest option, but it can be inconvenient if you are working in a team and need to share the state file.

- **Remote state file**: You can also store the state file in a remote location, such as an S3 bucket or a Terraform Cloud workspace. This allows you to share the state file with other members of your team, and it can also help protect against data loss if your local machine is destroyed.

- **Locking state file**: When using a remote state file, you can enable state file locking to prevent multiple users from modifying the state file simultaneously. This can help avoid conflicts when multiple users are making changes to the infrastructure at the same time.

Modules

Modules are self-contained packages of Terraform configurations that can be shared and reused. You can use modules to define your infrastructure in a more modular, reusable way.

For example, suppose you want to create a web server cluster with a load balancer and a database. You could create a module for each of these components, and then use those modules to build the web server cluster. This would allow you to reuse the modules in other infrastructure projects, and it would make your code easier to understand and maintain.

To create a module, you create a directory with one or more Terraform configuration files and a `module.tf` file that specifies the inputs and outputs of the module. You can then use the module in your infrastructure by calling it from another Terraform configuration using a module block.

Here's an example of a module block that calls a module named `web_server_cluster`:

```
module "web_server_cluster" {
  source = "./web_server_cluster"
  num_web_servers = 3
  web_server_size = "t2.micro"
  database_size = "t2.micro"
}
```

This block specifies the source of the module (a local directory named `web_server_cluster`) and sets the input variables for the module (`num_web_servers`, `web_server_size`, and `database_size`). The module can then use these input variables to create the desired infrastructure.

Variables

Variables allow you to parameterize your configurations and make them more flexible. You can use variables to define values that are used in multiple places in your configuration, or that you want to be able to adjust easily.

To define a variable in your Terraform configuration, you can use a variable block. Here's an example of a variable block that defines a variable named `image_id`:

```
variable "image_id" {
  type = string
}
```

This block defines a variable of type string named `image_id`. You can then use this variable in your configuration by referencing it with the `${var.name}` syntax, for example:

```
resource "aws_instance" "web_server" {
  ami           = "${var.image_id}"
  instance_type = "t2.micro"
}
```

You can also specify default values for your variables using the `default` attribute, for example:

```
variable "image_id" {
  type    = string
  default = "ami-12345678"
}
```

This sets the default value of the image_id variable to ami-12345678. If you don't specify a value for the variable when you run Terraform, it will use the default value.

You can set the value of a variable using one of the following methods:

- **Hard-coding the value**: You can set the value of a variable directly in your configuration using the default attribute. This is useful for simple configurations or for testing.

- **Using environment variables**: You can set the value of a variable using an environment variable. This can be useful for storing sensitive information or for managing multiple environments (e.g. staging and production).

- **Using a variable definition file**: You can store the values of your variables in a separate file (e.g. terraform.tfvars) and reference the file in your configuration using the -var-file flag. This can be useful for sharing values across multiple configurations or for storing values that you don't want to commit to version control.

Outputs

Outputs are values that are exported from a Terraform configuration and can be accessed from other configurations or consumed by external systems.

To define an output in your Terraform configuration, you can use an output block. Here's an example of an output block that exports the public IP address of an EC2 instance:

```
output "public_ip" {
    value = "${aws_instance.web_server.public_ip}"
}
```

This output block exports a value named public_ip and sets its value to the public IP address of the web_server EC2 instance. You can then access the value of this output using the terraform output command or by referencing it in another Terraform configuration using the ${output. name} syntax.

Outputs are useful for displaying important information about your infrastructure, such as the IP addresses of resources or the URLs of applications. They can also be used to pass information between multiple configurations, such as the ID of an S3 bucket that is created in one configuration and used in another.

Provisioners

Provisioners are used to execute scripts or make API calls after a resource is created. This can be useful for tasks such as installing software on an EC2 instance or uploading files to an S3 bucket.

Provisioners are defined in a `provisioner` block within a `resource` block. For example, the following `resource` block includes a provisioner that runs a shell script after the EC2 instance is created:

```
resource "aws_instance" "web_server" {
  ami           = "ami-12345678"
  instance_type = "t2.micro"

  provisioner "remote-exec" {
    inline = [
      "apt-get update",
      "apt-get install -y nginx",
    ]
  }
}
```

This provisioner uses the `remote-exec` provisioner type, which allows you to execute commands on the remote host using SSH. The inline attribute specifies the commands to run. In this case, the provisioner installs the `nginx` web server on the EC2 instance.

Terraform supports several different types of provisioners, including `file`, `remote-exec`, and `local-exec`. You can use different provisioners depending on your needs and the type of resource you are managing.

Selecting AWS Providers

In Terraform, a provider is a plugin that integrates Terraform with a specific infrastructure platform, such as AWS, Google Cloud, or Azure. Providers are responsible for understanding the API of the infrastructure platform and exposing resources that can be created, modified, and destroyed using Terraform.

There are two types of Terraform providers – official providers and third-party providers:

- **Official providers** are developed and maintained by HashiCorp, the company behind Terraform. These providers are considered the most stable and reliable, as they are supported by HashiCorp and receive regular updates.

- **Third-party providers** are developed and maintained by external organizations or individuals. These providers are not officially supported by HashiCorp, but they can be useful for extending Terraform to support additional infrastructure platforms or tools.

You can find a list of all the available Terraform providers on the Terraform Registry. (`https://registry.terraform.io/browse/providers`). The Registry lists both official and third-party providers, and it includes information on the providers' compatibility with Terraform and the resources they support.

To select a provider in Terraform, you need to specify it in a `provider` block in your configuration. For example, to use the AWS provider, you would add the following block to your configuration:

```
provider "aws" {
    region = "us-east-1"
}
```

This block specifies that you want to use the AWS provider and that you want to use the `us-east-1` region. You can also specify provider-specific configuration options, such as the access key and secret key to use when authenticating with the provider's APIs.

You can use multiple `provider` blocks in a single configuration to manage resources across multiple providers. For example, you might use the AWS provider to manage your EC2 instances and the Google Cloud provider to manage your Compute Engine instances.

To specify the version of a provider to use, you can use the `version` attribute in the `provider` block, for example:

```
provider "aws" {
    version = "~> 2.0"
}
```

This specifies that you want to use the latest version of the AWS provider that is compatible with version 2.0.

By running the `terraform init` command, Terraform will automatically download the required plugins for the providers specified in your configuration. To enforce a specific version of a provider, you can add a `version` attribute to the `provider` block.

You can also specify the version of a provider using the `-upgrade` flag when running `terraform init`. This will force Terraform to download the latest version of the provider, even if you have already downloaded a compatible version.

If you want to use a specific version of a provider, you can specify the version number in the `version` attribute, for example:

```
provider "aws" {
    version = "2.23.0"
}
```

This specifies that you want to use version 2.23.0 of the AWS provider.

If you want to use the latest version of a provider, you can set the `version` attribute to `latest`, for example:

```
provider "aws" {
  version = "latest"
}
```

This specifies that you want to use the latest version of the AWS provider that is available.

You can also use the `-get-plugins` flag when running `terraform init` to download the latest versions of all the providers in your configuration.

There are currently two AWS Providers available in Terraform:

- The **aws provider** is the legacy provider, and it is written in Go. It uses the AWS SDK for Go to make API requests to AWS, and it was the default AWS provider for Terraform for a long time.

- The **aws.sdk provider** is a newer provider, and it is written in TypeScript. It uses the AWS SDK for JavaScript to make API requests to AWS. It was introduced in Terraform v0.13 as an experimental provider, and it became a stable provider in Terraform v0.14.

In general, the `aws.sdk` provider is preferred over the `aws` provider, as it is more feature-complete and has better support for AWS services. However, the `aws` provider is still widely used and is likely to be supported for the foreseeable future.

Selecting AWS public modules for your needs

To use a public module in Terraform, you need to specify the source of the module in a `module` block in your configuration. You can specify the source of a module using a local path, Terraform Public Registry, a Git repository URL, or a URL to a compressed archive file (e.g. a `.zip` or `.tar.gz` file).

Here are some examples of how you can specify the source of a public module in a `module` block:

Local path: To use a module from a local directory, specify the path to the directory as the `source` attribute, for example:

```
module "web_server_cluster" {
  source = "./web_server_cluster"
}
```

This specifies that the `web_server_cluster` module is located in a local directory named `web_server_cluster`.

Git repository: To use a module from a Git repository, specify the URL of the repository as the `source` attribute, for example:

```
module "web_server_cluster" {
    source = "git::https://github.com/example/web_server_cluster.git"
}
```

This specifies that the `web_server_cluster` module is located in a Git repository at the URL `https://github.com/example/web_server_cluster.git`.

Compressed archive file: To use a module from a compressed archive file, specify the URL of the file as the `source` attribute, for example:

```
module "web_server_cluster" {
    source = "http://example.com/web_server_cluster.zip"
}
```

This specifies that the `web_server_cluster` module is located in a `.zip` file at the URL `http://example.com/web_server_cluster.zip`.

Terraform Registry: To use a module from the Terraform Registry in your configuration, you need to specify the module in a `module` block and set the `source` attribute to the URL of the module on the registry.

Here's an example of a `module` block that calls a module hosted on the Terraform Registry:

```
module "web_server_cluster" {
    source = "my-terraform-modules/web_server_cluster"
}
```

You can find a list of available public modules on the Terraform Registry (`https://registry.terraform.io/browse/modules`). The Registry includes both official and third-party modules, and it includes information on the module's compatibility with Terraform and the resources it supports. You can use the search bar to find specific modules, or you can browse the categories to find modules for a particular infrastructure platform or use case.

How to decide on Terraform module selection

There are several factors to consider when deciding which Terraform modules to use in your infrastructure:

- **Compatibility**: Make sure that the module is compatible with the version of Terraform you are using. You can check the compatibility information on the module's page on the Terraform Registry.

- **Supported resources**: Check that the module supports the resources you need to create or manage. You can find a list of the resources supported by the module on its page on the Terraform Registry.

- **Input variables**: Make sure that the module has the input variables you need to customize the behavior of the module. You can find a list of the input variables for the module on its page on the Terraform Registry.

- **Output values**: Check that the module has the output values you need to access the created resources or pass information between modules. You can find a list of the output values for the module on its page on the Terraform Registry.

- **Maintenance**: Consider the maintenance status of the module. Check when the module was last updated and whether it is actively maintained. You may want to choose a module that is more actively maintained to ensure that it stays up to date and receives bug fixes and new features.

How to write custom Terraform AWS modules

To write a Terraform module, you will need to create a configuration file or a set of configuration files that define the resources you want to create or manage. A module is essentially a reusable configuration that can be called from other configurations or modules.

Here are the steps you can follow to write a Terraform module:

1. **Define the resources you want to create or manage**: Use `resource` blocks to define the resources you want to create or manage. For example, you might use an `aws_instance` resource to create an EC2 instance or an `aws_s3_bucket` resource to create an S3 bucket.

2. **Define the input variables for the module**: Use `variable` blocks to define the input variables for the module. Input variables allow users of the module to customize the behavior of the module when it is called. For example, you might define a variable for the number of EC2 instances to create or the name of an S3 bucket.

3. **Define the output values for the module**: Use `output` blocks to define the output values for the module. Output values allow users of the module to access the created resources or pass information between modules. For example, you might define an output value for the public IP address of an EC2 instance or the URL of an S3 bucket.

4. **Test the module**: Use `terraform plan` and `terraform apply` to test the module and make sure it is creating the resources as expected.

5. **Document the module**: Document the input variables, output values, and any other important information about the module in the README file for the module.

Summary

In this chapter, we covered the basics of developing AWS infrastructure projects with Terraform. We learned how to deploy our first Terraform template in AWS, select AWS Providers, and choose public AWS modules that best suit our needs. We also explored how to write custom Terraform AWS modules and how to use them effectively. By the end of this chapter, you should be able to develop and deploy your own AWS infrastructure, make decisions about providers, select the best public AWS modules, and create and use custom Terraform AWS modules. These skills will serve as a foundation for you to continue exploring Terraform and AWS infrastructure development.

Whether you're new to serverless or an experienced developer, deploying serverless projects with Terraform can be a game changer for your development workflow. In the next chapter, we'll dive into the world of serverless and explore how to use Terraform to deploy serverless applications on AWS Lambda. From configuring AWS APIs and authentication to handling event triggers and scaling, you'll gain the skills needed to successfully deploy and manage your serverless projects with Terraform.

8

Deploying Serverless Projects with Terraform

Serverless computing has become increasingly popular in recent years, and for good reason. With AWS Lambda and AWS Fargate, you can develop and deploy your applications without the need to manage servers or infrastructure. Terraform makes it easy to design, deploy, and manage your serverless infrastructure on AWS.

In this chapter, we will explore the concepts of AWS landing zones and foundations, and how they can help you set up and manage your AWS accounts and infrastructure. We will cover the different options available for implementing landing zones and how to select the best design for your needs. Additionally, we will explore the use of AWS Organizations with Terraform to manage your AWS infrastructure.

Next, we will dive into the world of serverless computing, exploring what it is and when to use it. We will cover AWS Lambda and AWS Fargate, and how to use them to build and deploy your applications. We will also explore different deployment patterns and how to use Terraform to design and deploy your serverless infrastructure.

By the end of this chapter, you will have a solid understanding of AWS landing zones and foundations, as well as how to design and deploy serverless infrastructure using Terraform.

What are landing zones and why do we need them?

A landing zone is a reference architecture for a multi-account AWS environment. It provides a set of foundational resources and best practices that you can use as a starting point for your infrastructure.

A landing zone typically includes the following components:

- **A core account**: This is the primary account that contains the shared resources for the environment, such as the landing zone itself and the **Identity and Access Management** (**IAM**) resources.

- **One or more member accounts**: These are the accounts that contain the resources for your applications and workloads. The member accounts are linked to the core account and inherit the shared resources and policies from the core account.

- **A networking layer**: This includes the **Virtual Private Clouds** (**VPCs**) and other networking resources that are shared across the accounts.

- **A security layer**: This includes the IAM policies and other security resources that are shared across the accounts.

- **A governance layer**: This includes the policies and controls that are used to enforce compliance and manage the environment.

A landing zone can help you manage a multi-account environment more effectively and efficiently by providing a consistent set of resources and practices across all accounts. It can also help you onboard new accounts and applications more quickly by providing a standard framework to follow.

There are several reasons why you might want to use a landing zone in your AWS environment:

- **Improved security and compliance**: A landing zone provides a set of shared security and compliance resources, such as IAM policies and network controls, that are applied consistently across all accounts. This can help you improve the security and compliance of your environment by enforcing best practices and reducing the risk of misconfigurations.

- **Improved efficiency and automation**: A landing zone can help you automate the setup and management of your multi-account environment by providing a standard set of resources and practices to follow. This can save you time and effort and reduce the risk of errors.

- **Improved scalability and flexibility**: A landing zone can help you scale your environment more easily by providing a flexible, modular architecture that can accommodate new accounts and applications as needed.

- **Improved governance and control**: A landing zone can help you enforce governance and control over your environment by providing a central location for managing shared resources and policies.

AWS Foundations

AWS Foundations is a set of best practices and recommended configurations for building and managing infrastructure on AWS. It provides guidance on how to set up your AWS accounts, networking, security, and governance in a way that is scalable, secure, and compliant.

AWS Foundations includes recommendations for the following areas:

- **Account structure**: This includes guidance on how to set up and organize your AWS accounts and how to use AWS Organizations to manage them

- **Networking**: This includes guidance on how to set up your VPCs, subnets, and routing to create a scalable and secure network architecture

- **Security**: This includes guidance on how to secure your AWS resources using IAM, encryption, and other security controls

- **Governance**: This includes guidance on how to enforce compliance and manage your AWS environment using policies, controls, and monitoring

AWS Foundations is intended to provide a set of best practices and recommendations that you can use as a starting point for building and managing your infrastructure on AWS. It is not a one-size-fits-all solution, and you may need to tailor the recommendations to fit the specific needs of your organization.

AWS Foundations is not a product or a service offered by AWS, but rather a set of guidelines and recommendations that you can use to build your infrastructure on AWS.

AWS Foundations includes recommendations for setting up and organizing your AWS accounts, creating a scalable and secure networking architecture, securing your AWS resources, and enforcing compliance and governance in your environment.

AWS Foundations is intended to be a living document that is regularly updated with new best practices and recommendations as they become available.

How to build landing zones with Terraform in AWS

AWS Control Tower Account Factory is a feature of AWS Control Tower that allows you to automate the creation of member accounts in your multi-account AWS environment. With Account Factory, you can use Terraform templates to define the resources and configuration for your member accounts and then use the AWS Control Tower API to create the accounts and provision the resources automatically.

Here are some key features of AWS Control Tower Account Factory:

- **Automated account creation**: With Account Factory, you can use Terraform templates to define the resources and configuration for your member accounts and then use the AWS Control Tower API to create the accounts and provision the resources automatically.

- **Standardized account setup**: Account Factory allows you to enforce a standard set of resources and configuration for your member accounts, helping you ensure consistency and compliance across your environment.

- **Customization options**: You can use variables in your Terraform templates to customize the resources and configuration for your member accounts. This allows you to create accounts that are tailored to the specific needs of your organization.

- **Integration with AWS Control Tower**: Account Factory is integrated with AWS Control Tower, allowing you to use the AWS Control Tower dashboard to monitor and manage your member accounts.

There is an informative tutorial from Terraform on how to utilize Account Factory for Terraform (AFT): `https://developer.hashicorp.com/terraform/tutorials/aws/aws-control-tower-aft`.

What is serverless?

Serverless computing is a cloud computing execution model that allows the cloud provider to dynamically allocate resources to run the user's code, with the user only paying for the resources consumed. This model frees the user from the hassle of provisioning, scaling, and maintaining the underlying infrastructure.

In a serverless model, the user creates and deploys code in the form of functions, which are executed in response to events or invocations. The cloud provider automatically allocates the necessary resources to run the function, and the user only pays for the actual execution of the function.

Serverless computing can provide several benefits, including reduced operational overhead, scalability, and cost-efficiency. It is particularly well suited for applications that have intermittent or unpredictable workloads, as the user only pays for the resources used to run the code.

AWS offers several serverless computing services, including AWS Lambda, which allows you to run code in response to events or invocations, and AWS Fargate, which allows you to run containerized applications in a serverless environment.

What are AWS serverless patterns?

AWS serverless patterns are templates for common ways to build and deploy applications using serverless technologies on **Amazon Web Services** (**AWS**). These patterns provide guidance on how to design and architect your applications to take advantage of the benefits of serverless computing.

There are many different serverless patterns that you can use, depending on the requirements of your application. Some common serverless patterns include the following:

- **Event-driven architecture**: This pattern involves building applications that respond to events, such as a user uploading a file or a sensor sending data

- **Microservices**: This pattern involves breaking up a large application into smaller, independent services that can be developed, deployed, and scaled independently

- **Data processing**: This pattern involves using serverless technologies to process large amounts of data, such as converting data from one format to another or aggregating data from multiple sources
- **Web applications**: This pattern involves using serverless technologies to build and deploy web applications, such as static websites or dynamic web applications

These are just a few examples of the types of serverless patterns that are available. There are many other patterns that you can use to build and deploy your applications on AWS.

AWS serverless resources are resources that are used to build and deploy applications using serverless technologies on AWS. These resources can include a variety of different services and tools, such as the following:

- **AWS Lambda**: A service that lets you run code without provisioning or managing servers
- **Amazon API Gateway**: A service that makes it easy to create, publish, maintain, monitor, and secure APIs
- **AWS Fargate**: A service that lets you run containers without having to manage the underlying EC2 instances
- **AWS Step Functions**: A service that makes it easy to coordinate the functions of distributed applications and microservices
- **AWS App Runner**: A service that makes it easy to build and deploy containerized applications

What is AWS Lambda?

AWS Lambda is a fully managed serverless compute service that allows you to execute your code in response to various events, such as changes to data in an Amazon S3 bucket or a new item being added to a DynamoDB table. It automatically manages the underlying compute resources for you, so you don't need to worry about provisioning or maintaining any servers. This enables you to focus on writing and deploying your code, without any administrative overhead.

Some common use cases for AWS Lambda include the following:

- Running backend logic for web and mobile applications
- Processing data streams and event triggers
- Automating maintenance and administration tasks

In Lambda, you write your code and then upload it to the service. When an event occurs that triggers your code, Lambda executes it and automatically scales the underlying infrastructure to run your code. You only pay for the compute time that you consume, so you can run your code without having to worry about managing servers or infrastructure.

AWS Lambda supports a variety of programming languages, including Node.js, Python, Java, C#, and Go, and you can use it in conjunction with other AWS services to build powerful and scalable applications.

Let's look at some key features:

- **Lambda functions**: In Lambda, you create functions that contain your code. These functions are triggered by events, such as a user uploading a file to Amazon S3 or a request to an API Gateway endpoint. You can specify the events that trigger your functions, and Lambda automatically executes the function when these events occur.

- **Execution environment**: AWS Lambda provides a fully managed execution environment for your functions. This includes the infrastructure and operating system, as well as the language runtime (e.g., Node.js, Python, Java, etc.). When you create a function, you can specify the runtime and the amount of memory that you want to allocate to your function.

- **Scaling**: One of the benefits of using AWS Lambda is that it automatically scales to meet the needs of your application. When your function is invoked, Lambda allocates the necessary compute resources to run your code. If your function is invoked more frequently, Lambda automatically scales up to meet the increased demand.

- **Integrations**: AWS Lambda integrates with a wide variety of other AWS services, allowing you to build powerful and scalable applications. For example, you can use Lambda with Amazon S3 to automatically process files as they are uploaded to the bucket, or with Amazon DynamoDB to automatically update records in the database as they are added or modified.

What is AWS Fargate?

AWS Fargate is a fully managed service that makes it easier to run containerized applications on AWS. AWS Fargate removes the need to manage the underlying infrastructure, so you can focus on building and running your applications.

With AWS Fargate, you simply specify the number and type of resources that you want to allocate to your applications, and AWS Fargate takes care of the rest. It automatically allocates the necessary compute resources, such as Amazon **Elastic Compute Cloud** (**EC2**) instances, and ensures that your containers are running in a highly available and scalable manner.

AWS Fargate is a good choice for developers who want to run containerized applications on AWS without the overhead of managing the underlying infrastructure. It is especially well suited for applications that require rapid scaling or that have unpredictable workloads, as AWS Fargate can automatically scale your resources up or down as needed.

AWS Fargate is a fully managed service, which means that AWS takes care of the underlying infrastructure for you. This includes provisioning and managing the EC2 instances that run your tasks, as well as handling any infrastructure maintenance or patching.

AWS Fargate is available in all regions where Amazon ECS is available, and you can use it to run tasks on both Amazon ECS and Amazon **Elastic Kubernetes Service (EKS)**.

AWS Fargate supports all of the same features as Amazon ECS, including the ability to use Amazon ECS task definitions to define your tasks, integration with other AWS services such as Amazon CloudWatch and AWS IAM, and support for Docker containers.

AWS Fargate is well suited for use cases where you want to run containers but don't want to worry about the underlying infrastructure. It can be a good choice for developers who want to focus on building and deploying applications rather than managing infrastructure, or for organizations that want to run containerized applications but don't have the in-house expertise to manage the underlying infrastructure.

You can use AWS Fargate with Amazon **Elastic Container Service (ECS)** or Amazon EKS to run your containerized applications. It is also integrated with other AWS services, such as Amazon CloudWatch and AWS IAM, which you can use to monitor and secure your applications.

How to design a serverless infrastructure with Terraform

Here are general steps you can follow to design a serverless infrastructure using Terraform:

1. Identify the components of your infrastructure that can be implemented as serverless resources. This might include things such as APIs, backend workers, and data processing pipelines.

2. Determine which serverless platforms and services you will use to implement these components. This might include services such as AWS Lambda, Amazon API Gateway, and Amazon DynamoDB, or managed services such as AWS Fargate or AWS AppSync.

3. Define the required IAM roles and permissions for your serverless resources. This will typically involve creating IAM policies and attaching them to IAM roles that your resources can assume.

4. Use Terraform to create the necessary infrastructure resources, such as VPCs, security groups, and subnets. You can also use Terraform to create and configure the serverless resources themselves.

5. Define any dependencies between your resources using Terraform's dependencies syntax. This will ensure that resources are created in the correct order and that any necessary connections are established between them.

6. Use Terraform's testing and validation features to ensure that your infrastructure is configured correctly and adheres to best practices. This might include things such as running `terraform plan` to preview changes, or running `terraform validate` to check for syntax errors.

7. Use Terraform's version control integration to manage changes to your infrastructure over time. This will allow you to track changes, roll back to previous versions if necessary, and collaborate with other team members.

Moreover, here are some additional points to consider for designing the serverless infrastructure:

- Decide on a deployment strategy that works for your organization. This might include using Terraform's `apply` command to deploy changes directly or using a **continuous integration/ continuous deployment** (**CI/CD**) platform such as AWS CodePipeline to automate the deployment process.

- Consider the scalability and availability requirements of your serverless resources. You can use Terraform to specify things like the number of replicas for an Amazon ECS service, or the number of function instances for an AWS Lambda function.

- Use Terraform's output values to expose important information about your infrastructure to other tools and processes. For example, you might output the URL of an API Gateway endpoint so that it can be used by other parts of your infrastructure.

- Use Terraform's workspaces feature to manage multiple environments, such as production, staging, and development. This will allow you to easily switch between environments and apply changes to the appropriate environment.

- Consider using Terraform modules to encapsulate reusable pieces of infrastructure. This can help you reduce duplication and make it easier to manage and maintain your infrastructure over time.

How to develop a serverless infrastructure

To develop a serverless infrastructure, you can follow these general steps:

1. Identify the components of your infrastructure that can be implemented as serverless resources. This might include things such as APIs, backend workers, and data processing pipelines.

2. Determine which serverless platforms and services you will use to implement these components. This might include services such as AWS Lambda, Amazon API Gateway, and Amazon DynamoDB, or managed services such as AWS Fargate or AWS AppSync.

3. Define the required IAM roles and permissions for your serverless resources. This will typically involve creating IAM policies and attaching them to IAM roles that your resources can assume.

4. Use the relevant tools and APIs to create and configure your serverless resources. This might include using the AWS Management Console, the AWS CLI, or the AWS SDKs.

5. Define any dependencies between your resources, such as connections between an API Gateway and a Lambda function, or between a Lambda function and a DynamoDB table.

6. Test your infrastructure to confirm that everything is working as expected.

7. Use monitoring and logging tools to track the performance and health of your serverless resources. This might include using Amazon CloudWatch to monitor resource metrics and logs or using AWS X-Ray to trace requests as they flow through your infrastructure.

How to deploy a serverless infrastructure using Terraform

To deploy a serverless infrastructure using Terraform, you can follow these steps:

1. Write Terraform configuration files to define the desired state of your infrastructure. These configuration files can use the **HashiCorp Configuration Language** (**HCL**) to specify the resources that you want to create, as well as the properties of those resources.

2. If you are following an event-driven architecture, you should consider carving all triggers and resources into the same Terraform project.

3. Use the `terraform init` command to initialize your working directory and download any necessary plugins or dependencies.

4. Use the `terraform plan` command to preview the changes that Terraform will make to your infrastructure. This will allow you to see what resources will be created, modified, or destroyed, and to confirm that the changes are what you expect.

5. Use the `terraform apply` command to apply the changes to your infrastructure. This will create or update the resources according to your configuration files.

6. Test your infrastructure to confirm that everything is working as expected.

7. Use Terraform's version control integration to manage changes to your infrastructure over time. This will allow you to track changes, roll back to previous versions if necessary, and collaborate with other team members.

8. Create a separate S3 bucket with relevant permissions to move your Terraform state file to secure and make it easy for collaboration with other team members.

9. Consider creating a pipeline in your CI/CD system to execute your Terraform templates for security and observability.

10. Avoid provisioning manual resources; utilize Terraform to cover all your resources, configurations, and environments. Any existing or legacy resources can easily be imported to Terraform.

Summary

In this chapter, we learned about deploying serverless projects with Terraform. We covered the basics of serverless computing, AWS Lambda, and AWS Fargate, and how to design and deploy serverless infrastructure with Terraform. We also explored the importance of AWS landing zones and how to select and implement them. Additionally, we discussed AWS Organizations and how to use them with Terraform.

In the next chapter, we will explore deploying containers in AWS with Terraform. We will cover the basics of containers, AWS ECS, Amazon EKS, and how to deploy containers with Terraform. We will also discuss best practices for deploying containers in AWS and how to use Terraform to manage container deployments.

9

Deploying Containers in AWS with Terraform

In recent years, containerization has become an increasingly popular method for deploying and managing applications in the cloud. **Amazon Web Services (AWS)** offers a range of containerization services, including Amazon **Elastic Container Registry (ECR)**, Amazon **Elastic Container Service (ECS)**, and Amazon **Elastic Kubernetes Service (EKS)**. In this chapter, you will learn how to use Terraform to deploy containers in AWS, from selecting and designing the appropriate infrastructure to developing and deploying your container infrastructure.

Get ready to dive into the world of containerization and learn how to deploy containers in AWS using Terraform with the following topics:

- What are containers?
- AWS containers
- How to utilize Terraform for containers
- How to use Terraform for AWS container resources

What are containers?

Containers are a type of virtualization technology that allows developers to package up an application and its dependencies into a single container, which can be easily moved between different environments. Containers provide a consistent environment for the application to run in, regardless of the underlying infrastructure. They are lightweight and efficient, as they share the host operating system kernel and do not require a full **virtual machine (VM)**. Popular containerization platforms include Docker and Kubernetes.

Containers offer a more lightweight and efficient alternative to VMs. In essence, a container is a self-contained, portable, and executable package that contains all the necessary components to run specific software, such as the code, runtime, libraries, environment variables, and configuration files. Because containers provide a consistent environment for the application to run in, they are well suited for use in various environments, including development, testing, and production.

Containers are built on top of a container engine, such as Docker or **Linux Containers** (**LXC**). These engines provide an abstraction layer on top of the host operating system and manage the container's resources, such as CPU, memory, and storage. Containers can be run on a single host or can be orchestrated across multiple hosts using container orchestration platforms such as Kubernetes, Amazon EKS, Amazon ECS, or Docker Swarm.

Containers are also highly portable, so they can be easily moved between different environments, such as from a developer's laptop to a test environment and then to production. This makes it easier to manage the entire application life cycle and ensures consistency across different stages of development.

In summary, containers are a way to package software in a format that can run consistently across different environments. They are lightweight, efficient, and easy to manage, making them a popular choice for modern application development and deployment.

Containers in AWS

In AWS, containers refer to a way of packaging and deploying applications as container images. These container images can be run on AWS services such as Amazon ECS and Amazon EKS.

Amazon ECS is a fully managed container orchestration service that makes it easy to run, scale, and secure containerized applications. With ECS, you can run containers on a cluster of Amazon **Elastic Compute Cloud** (**EC2**) instances, and it automatically handles tasks such as scaling, load balancing, and health monitoring. ECS also integrates with other AWS services such as **Elastic Load Balancing** (**ELB**), Amazon **Relational Database Service** (**RDS**), and Amazon **Simple Storage Service** (**S3**).

Amazon EKS is a managed service that makes it easy to deploy, scale, and operate containerized applications using Kubernetes. EKS automates the provisioning and management of the Kubernetes control plane and worker nodes, so you can focus on building and running your applications. EKS also integrates with other AWS services, such as ELB and Amazon RDS, to provide a fully managed Kubernetes experience on AWS.

AWS also offers other services that can be used in conjunction with containers, such as Amazon ECR for storing and managing container images, and AWS Fargate for running containers without the need to manage the underlying infrastructure.

In summary, in AWS, containers refer to containerized applications that can be run and managed on AWS services such as ECS and EKS, and other related services such as ECR and Fargate that provide a fully managed container orchestration service, allowing developers to focus on building and running their applications without worrying about the underlying infrastructure.

The reasons for using containers

There are several reasons why developers and organizations use containers:

- **Portability**: Containers provide a consistent environment for an application to run in, regardless of the underlying infrastructure. This makes them highly portable, so they can be easily moved between different environments such as development, testing, and production.

- **Isolation**: Containers provide isolation between different applications running on the same host, which helps to prevent conflicts and ensures that each application has the resources it needs.

- **Scalability**: Containers can be easily scaled up or down to meet changing demands, allowing for more efficient use of resources.

- **Cost-effective**: Containers are lightweight and share the host operating system kernel, so they are more efficient than full VMs. This means that you can run more containers on a single host, which can help to reduce costs.

- **Automation**: Containers can be easily automated and orchestrated using tools such as Kubernetes and Docker, making it easier to manage the entire application life cycle.

- **Efficiency**: Containers can be built and deployed faster, leading to faster development cycles and faster **time-to-market** (**TTM**).

- **Security**: Containers provide an additional layer of security by isolating the application from the host operating system and other applications running on the same host.

- **Microservices**: Containers can be used to deploy microservices-based architectures, which can make it easier to build and maintain complex applications.

- **Flexibility**: Containers can be used with a variety of platforms and technologies, such as Linux, Windows, and cloud providers, making them a flexible choice for different types of applications and environments.

- **Versioning**: Containers can be versioned, making it easy to roll back to a previous version of an application if necessary.

- **Testability**: Containers make it easy to test applications in different environments, as the entire application and its dependencies are packaged together in a container.

- **Continuous integration and deployment**: Containers can be integrated with **Continuous Integration and Continuous Deployment** (**CI/CD**) pipelines, allowing for automated building, testing, and deployment of applications.

- **Hybrid and multi-cloud**: Containers can be used to deploy and run applications across multiple cloud providers, allowing for greater flexibility and choice when it comes to cloud infrastructure.

- **Serverless**: Containers can be used in conjunction with serverless platforms such as AWS Lambda, Azure Functions, and Google Cloud Functions, to create highly scalable, event-driven applications.

In summary, containers provide a consistent and isolated environment, helping to ensure that an application will run the same way across different environments and making it easy to move the application between different environments. They are lightweight, easy to automate and scale, cost-effective, efficient, and provide additional security. They are also a good fit for microservices-based architectures.

How to containerize applications

There are several steps involved in containerizing an application:

1. **Package the application and its dependencies**: The first step is to package the application and its dependencies into a single container. This typically involves creating a container image, which includes the application code, runtime, libraries, environment variables, and config files.

2. **Define the container's environment**: The next step is to define the container's environment, including the operating system and runtime that the application will run on. This is done by creating a Dockerfile, which specifies the base image to use, any additional software to install, and any configuration settings.

3. **Build the container image**: Once the Dockerfile is defined, the container image can be built using a tool such as Docker. This creates a lightweight, standalone, executable package that includes everything needed to run the application.

4. **Push the container image to a registry**: After the container image is built, it can be pushed to a container registry, such as Docker Hub or Amazon ECR, where it can be easily shared and distributed to different environments.

5. **Deploy the container**: The final step is to deploy the container to a container orchestration platform, such as Kubernetes or Amazon ECS, where it can be easily scaled and managed.

6. **Test the containerized application**: Before deploying the containerized application to production, it's important to test it in a non-production environment to make sure it works as expected. This can be done by running the container image on a test cluster or on a developer's local machine. This step can help identify and fix any issues before the application is deployed to production.

7. **Optimize the container image**: It's important to optimize the container image to minimize the size and reduce the number of layers. This can be done by using multi-stage builds, removing unnecessary files and packages, and using smaller base images.

8. **Monitor and update the containerized application**: Once the containerized application is deployed, it's important to monitor it to ensure it's running smoothly and to identify any potential issues. Regular updates and security patches should be applied to the containerized application and its dependencies.

9. **Consider security best practices**: Security should always be considered when containerizing an application. Best practices include running containers with the least privilege, using a container registry with built-in security features, and regularly updating the container images and the host system.

In summary, containerizing an application is a multi-step process that involves packaging the application and its dependencies into a container image, defining the container's environment, building the image, pushing it to a registry, deploying it to a container orchestration platform, testing it, optimizing the image, monitoring and updating the application, and considering security best practices.

AWS containers

In AWS, containers refer to a way of packaging and deploying applications as container images. These container images can be run on AWS services such as Amazon ECS and Amazon EKS.

Amazon ECS and Amazon EKS are explained in the *Containers in AWS* section, so we won't repeat them here.

AWS Fargate is a serverless compute engine for containers that allows you to run containers without having to provision and manage the underlying infrastructure. With Fargate, you only pay for the resources that your containers use, and there is no need to manage the underlying EC2 instances.

Amazon ECR is a fully managed container registry service that makes it easy to store, manage, and deploy container images. ECR is integrated with other AWS services such as ECS and EKS, making it easy to store and retrieve container images for use in those services.

AWS App Runner is a fully managed service that makes it easy to build, test, and deploy containerized applications quickly. It automates the building, testing, and deployment of containerized applications, allowing developers to focus on writing code.

AWS Elastic Beanstalk is a fully managed service that makes it easy to deploy, run, and scale web applications and services. Elastic Beanstalk supports multiple platforms, including Java, .NET, PHP, Node.js, Python, Ruby, and Go, and it also supports deploying applications as Docker containers.

AWS Lambda is a serverless compute service that allows you to run code without provisioning or managing servers. It automatically scales your applications in response to incoming requests, and you only pay for the compute time that you consume. AWS Lambda with container support allows developers to package their application code and dependencies together in a container and deploy it as a function. This enables developers to take advantage of the benefits of containers such as consistent runtime environments and the ability to run their applications in different environments.

In summary, AWS offers a range of services that can be used to deploy and manage containerized applications, including Amazon ECS, Amazon EKS, AWS Fargate, Amazon ECR, and AWS App Runner. These services provide an easy way to deploy, run, and manage containerized applications, integrate with other AWS services, and automate various aspects of the application life cycle management.

How to choose the best containerization platform in AWS

Choosing the best containerization platform in AWS will depend on the specific requirements of your application and use case. Here are some factors to consider when making your decision:

- **Microservices versus monolithic**: If your application is built using a microservices-based architecture, then ECS or EKS would be a good choice, as they are designed to handle the scaling and orchestration of multiple services. If your application is a monolithic application, then Fargate or App Runner may be a better fit.

- **Scale**: Consider the scale of your application and the resources it requires. ECS and EKS are both highly scalable and can handle large numbers of containers and services. Fargate is also scalable, but it is more suited for running small to medium-sized applications.

- **Existing infrastructure**: If you already have an existing infrastructure in place, it may be more cost-effective to use ECS or EKS, as they can integrate with your existing resources.

- **Cost**: Consider the cost of running your application on each platform. ECS and EKS may be more expensive than Fargate, as they require the provisioning and management of underlying infrastructure.

- **Functionality**: Consider the functionality that you need for your application. ECS and EKS provide more advanced features for deploying, scaling, and managing containerized applications, while Fargate is more suited for running individual containers.

- **Team experience**: Consider the experience of your team with the different platforms. If your team has experience with Kubernetes, EKS might be a better fit; if it has experience with AWS native services, ECS or Fargate might be more appropriate.

Each platform has its own set of features and capabilities, and the choice of which platform to use will depend on the specific requirements of your application and use case. ECS and EKS are more suited for microservices-based architectures, while Fargate and App Runner are more suited for running individual containers. AWS Lambda is more suited for running function-based workloads, and Elastic Beanstalk is more suited for deploying web applications and services.

Ultimately, the best containerization platform for your application will depend on the specific requirements of your use case. It is important to evaluate each platform based on the factors that are most important to your application, such as scalability, cost, and functionality. It may also be beneficial to test different platforms in a non-production environment to determine which one works best for your application before making a final decision.

Additionally, it's worth considering the level of flexibility and control you want over the infrastructure and the level of automation you want to achieve. ECS and EKS provide more control and flexibility over the infrastructure, while Fargate and App Runner provide more automation.

In general, it's recommended to start with the simplest option that meets your needs and gradually add complexity as needed. AWS Lambda, for instance, is a good starting point for function-based workloads, Elastic Beanstalk for web-based applications, Fargate for small to medium-sized applications, and ECS or EKS for complex microservices-based architectures.

It's also important to note that AWS provides a variety of services that can be used in conjunction with the containerization platform, such as ECR for storing and managing container images, and AWS App Mesh for service mesh management.

How to utilize Terraform for containers

Terraform provides a powerful platform for managing and deploying container infrastructure on AWS. With Terraform, you can easily create and manage resources such as ECR, ECS, and EKS. This section will cover the basics of how to utilize Terraform for containers, including selecting and designing container infrastructure with Terraform, and how to develop and deploy container infrastructure using Terraform.

Deploying containers with Terraform

Terraform is a tool that allows you to define, provision, and manage infrastructure as code. To design a container using Terraform, you can use the `docker_container` resource to create, configure, and manage a container.

Here is an example of how to use Terraform to create a container:

```
resource "docker_container" "example" {
  name  = "example-container"
  image = "nginx:latest"

  ports {
    internal = 80
    external = 8080
  }

  environment {
    EXAMPLE_VAR = "example value"
  }

  volumes {
    container_path = "/var/www/html"
    host_path = "./data"
    read_only = true
  }
}
```

This example creates a container named `"example-container"` using the latest version of the nginx image, maps port 80 inside the container to port 8080 on the host, and sets an environment variable named EXAMPLE_VAR with a value of `"example value"`. The container also creates a volume that maps the `/var/www/html` path inside the container to the `./data` path on the host, with read-only access.

You can also use the `docker_image` resource to create, manage, and configure a container image, and the `docker_network` resource to create, manage, and configure container networks.

Here is an example of how to use Terraform to create a container image:

```
resource "docker_image" "example" {
  name = "example-image"
  build {
    context = "./example-image"
    dockerfile = "Dockerfile"
  }
}
```

This example creates a container image named `"example-image"` using the Dockerfile in the `"./example-image"` directory.

Here is an example of how to use Terraform to create a container network:

```
resource "docker_network" "example" {
  name = "example-network"
  driver = "bridge"
}
```

This example creates a container network named `"example-network"` with a `bridge` driver.

By using the `docker_container`, `docker_image`, and `docker_network` resources, you can use Terraform to create, manage, and configure containers, container images, and container networks in a repeatable and automated way.

Terraform also supports other providers besides Docker, such as AWS ECS, ECR, and EKS, **Azure Container Instance** (**ACI**), and Google Container Engine, which provides more specific resources and data sources that are tailored to those specific providers.

How to use Terraform for AWS container resources

There are several ways to deploy containers in AWS, depending on your specific requirements and use case. Here are the general steps to deploy a container in AWS:

1. Build and push your container image to a container registry such as Amazon ECR or any other public or private registry

2. Choose a container orchestration platform such as Amazon ECS, Amazon EKS, AWS Fargate, AWS Lambda, AWS Elastic Beanstalk, or AWS App Runner

3. Create a task definition or Pod definition that describes the container image and its configurations, such as environment variables, ports, and volumes

4. Create a service or deployment that uses the task definition or Pod definition to launch one or more instances of the container

5. Optionally, configure scaling, load balancing, and monitoring for your containerized application

6. Optionally, you can use services such as Terraform or AWS CloudFormation to automate the deployment and management of your container infrastructure

7. Test your application and monitor its performance to make sure it's working as expected

It's worth noting that each of the container orchestration platforms that AWS provides has its own set of management consoles, APIs, and CLIs that you can use to deploy, manage, and scale your containerized applications.

After building container images to push container images to ECR, we can utilize Terraform to create ECR repositories.

How to deploy AWS ECR with Terraform

Amazon ECR is a fully managed container registry service that makes it easy to store, manage, and deploy container images. To use Terraform to manage ECR resources and to deploy an Amazon ECR repository, you can use the AWS provider for Terraform, which provides a set of resources and data sources specific to ECR. Here are the general steps to deploy an ECR repository using Terraform:

1. Install and configure the AWS provider for Terraform in your local environment

2. Create a new Terraform configuration file and specify the AWS provider and the `aws_ecr_repository` resource

3. Define the properties of the ECR repository, such as the repository name, in the resource configuration

4. Run `terraform init` to initialize the Terraform environment and download the necessary provider plugins

5. Run `terraform plan` to preview the changes that will be made to your infrastructure

6. Run `terraform apply` to create an ECR repository in your AWS account

Here is an example of how to use Terraform to create an ECR repository:

```
provider "aws" {
  region = "us-west-2"
}


resource "aws_ecr_repository" "example" {
  name = "example-repository"
}
```

This example creates an ECR repository named `"example-repository"` in the `"us-west-2"` region.

You can also use the `aws_ecr_lifecycle_policy` resource to manage the life cycle policies for an ECR repository and the `aws_ecr_image` resource to manage the images stored in an ECR repository using Terraform.

Here is an example of how to use Terraform to create a life cycle policy for an ECR repository:

```
resource "aws_ecr_lifecycle_policy" "example" {
  repository = aws_ecr_repository.example.name
  policy = <<EOF
  {
    "rules": [
      {
        "rulePriority": 1,
        "description": "Expire images older than 30 days",
        "selection": {
          "tagStatus": "untagged",
          "countType": "sinceImagePushed",
          "countUnit": "days",
          "countNumber": 30
        },
        "action": {
          "type": "expire"
        }
      }
    ]
  }
  EOF
}
```

This example creates a life cycle policy for the ECR repository specified by the `aws_ecr_repository.example.name` reference. This policy expires images that are older than 30 days and have no tag associated with them.

It's important to note that this is a simple example of a life cycle policy. You can use the full set of options that an AWS ECR life cycle policy provides to create more complex policies, such as image tagging rules, image scanning rules, and so on.

You can also use the `terraform plan` and `terraform apply` commands to preview and apply the changes made to the repository policy.

Here is an example of how to use Terraform to create an image in an ECR repository:

```
resource "aws_ecr_image" "example" {
  repository = aws_ecr_repository.example.name
  image_tag = "latest"
  image_digest = "${data.aws_ecr_image.example.image_digest}"
}

data "aws_ecr_image" "example" {
  repository = aws_ecr_repository.example.name
  image_tag = "latest"
}
```

This example creates an image in an ECR repository specified by the `aws_ecr_repository.example.name` reference. The image is tagged with `"latest"` and the digest of the image is obtained from the `aws_ecr_image` data source.

You can use the `aws_ecr_image` resource to push and pull images to and from an ECR repository, as well as to manage images stored in an ECR repository.

The `aws_ecr_image` resource also allows you to specify image details such as the image tag, image digest, image manifest, and image scanned status.

It's important to note that the preceding example is a simple example of creating an image in an ECR repository. You can use the full set of options that the `aws_ecr_image` resource provides to create and manage images in your ECR repository.

Deploying container images to AWS container platforms with Terraform

In this section, we will explore how to deploy container images to AWS container platforms using Terraform. By utilizing Terraform, we can simplify the process of managing container infrastructure and automate the deployment of containerized applications on AWS. We will discuss the use of AWS container services such as ECR, ECS, and EKS and how to deploy container images to these services using Terraform.

Deploying to AWS ECS

To deploy container images to Amazon ECS using Terraform, you can use the AWS provider for Terraform, which provides a set of resources and data sources specific to ECS. Here are the general steps to deploy an ECS container using Terraform:

1. Install and configure the AWS provider for Terraform in your local environment.

2. Create a new Terraform configuration file and specify the AWS provider and the necessary ECS resources such as `aws_ecs_task_definition`, `aws_ecs_service`, and `aws_ecs_cluster`.

3. Define the properties of the container, such as the container image, container name, port mappings, and environment variables, in the task definition resource.

4. Create a service resource that references the task definition, and configure the desired number of task replicas and the load balancer settings if applicable.

5. Create a cluster resource if it doesn't exist, and reference it in the task definition and service resources.

6. Run `terraform init` to initialize the Terraform environment and download the necessary provider plugins.

7. Run `terraform plan` to preview the changes that will be made to your infrastructure.

8. Run `terraform apply` to create the ECS service and deploy the container to your cluster in your AWS account.

Here is an example of how to use Terraform to deploy a container to ECS:

```
resource "aws_ecs_task_definition" "example" {
  family = "example-task-definition"
  container_definitions = <<DEFINITION
[
  {
    "name": "example-container",
    "image": "example-image:latest",
    "portMappings": [
      {
        "containerPort": 80,
        "hostPort": 80
      }
    ],
    "memory": 512,
    "cpu": 256
  }
]
```

```
DEFINITION
}

resource "aws_ecs_service" "example" {
  name             = "example-service"
  task_definition  = aws_ecs_task_definition.example.arn
  cluster          = aws_ecs_cluster.example.id
  desired_count    = 2
}

resource "aws_ecs_cluster" "example" {
  name = "example-cluster"
}
```

This example creates an ECS task definition, service, and cluster using Terraform. The task definition defines the container image, container name, port mappings, and memory and CPU requirements. The service references the task definition and creates two replicas of the container in the specified ECS cluster.

You can also use the `aws_elbv2_listener` and `aws_elbv2_target_group` resources to configure a load balancer and register the ECS service as a target group.

It's worth noting that this is a simple example of deploying an ECS container using Terraform. You can use the full set of options that the ECS resources provide to create and manage more complex ECS environments, such as auto scaling, rolling updates, and integration with other AWS services such as CloudWatch, CloudTrail, and more.

You can also use the `terraform plan` and `terraform apply` commands to preview and apply the changes made to the ECS environment.

Deploying to AWS EKS

There are two steps for deploying applications to AWS EKS, as detailed next.

Creating an AWS EKS cluster with Terraform

To create an Amazon EKS cluster using Terraform, you can use the AWS provider for Terraform, which provides a set of resources and data sources specific to EKS. Here are the general steps to create an EKS cluster using Terraform:

1. Install and configure the AWS provider for Terraform in your local environment.
2. Create a new Terraform configuration file and specify the AWS provider and the `aws_eks_cluster` resource.

3. Define the properties of the EKS cluster, such as the cluster name, Kubernetes version, and VPC settings, in the resource configuration.

4. Optionally, create an **Identity and Access Management (IAM)** role and policy for the cluster, and associate them with the `aws_eks_cluster` resource.

5. Optionally, create a configuration file for `kubeconfig` to use the cluster.

6. Run `terraform init` to initialize the Terraform environment and download the necessary provider plugins.

7. Run `terraform plan` to preview the changes that will be made to your infrastructure.

8. Run `terraform apply` to create the EKS cluster in your AWS account.

Here is an example of how to use Terraform to create an EKS cluster:

```
resource "aws_eks_cluster" "example" {
  name      = "example-cluster"
  role_arn = aws_iam_role.example.arn
  version   = "1.20"
  vpc_config {
    security_group_ids = [aws_security_group.example.id]
    subnet_ids         = [aws_subnet.example.*.id]
  }
}

resource "aws_iam_role" "example" {
  name = "example-role"

  assume_role_policy = <<EOF
{
  "Version": "2012-10-17",
  "Statement": [
    {
      "Effect": "Allow",
      "Principal": {
        "Service": "eks.amazonaws.com"
      },
      "Action": "sts:AssumeRole"
    }
  ]
}
EOF
}

resource "aws_iam_role_policy" "example" {
```

```
    name = "example-policy"
    role = aws_iam_role.example.id

    policy = <<EOF
{
  "Version": "2012-10-17",
  "Statement": [
    {
      "Effect": "Allow",
      "Action": [
        "ec2:Describe*",
       "iam:PassRole",
"eks:"
],
"Resource": ""
}
]
}
EOF
}

resource "aws_security_group" "example" {
name = "example-security-group"
description = "Controls access to the EKS cluster"
}

resource "aws_subnet" "example" {
count = 2

vpc_id = aws_vpc.example.id
cidr_block = "10.0.${count.index}.0/24"
availability_zone = "us-west-2a"
map_public_ip_on_launch = true
}
```

This example creates an EKS cluster with the specified name and Kubernetes version and associates it with the specified IAM role and security group.

It also creates two subnets in the specified Availability Zone for worker nodes to launch into.

It's worth noting that this is a simple example of creating an EKS cluster using Terraform. You can use the full set of options that the EKS resources provide to create and manage more complex EKS environments, such as scaling, monitoring, and integration with other AWS services such as CloudWatch, CloudTrail, and more.

You can also use the `terraform plan` and `terraform apply` commands to preview and apply the changes made to the EKS environment.

Deploying an application to an AWS EKS cluster with Terraform

To deploy container images to Amazon EKS using Terraform, you can use the Kubernetes provider for Terraform, which provides a set of resources and data sources specific to EKS. Here are the general steps to deploy a Kubernetes Pod using Terraform:

1. Install and configure the Kubernetes provider for Terraform in your local environment.
2. Create a new Terraform configuration file and specify the Kubernetes provider and the necessary resources such as `kubernetes_namespace`, `kubernetes_deployment`, and `kubernetes_service`.
3. Define the properties of the Pod, such as the container image, container name, container ports, and environment variables, in the deployment resource.
4. Create a service resource that references the deployment, and configure the load balancer settings if applicable.
5. Create a namespace resource if it doesn't exist, and reference it in the deployment and service resources.
6. Run `terraform init` to initialize the Terraform environment and download the necessary provider plugins.
7. Run `terraform plan` to preview the changes that will be made to your infrastructure.
8. Run `terraform apply` to create the Kubernetes deployment and service, and deploy the Pod to your EKS cluster.

Here is an example of how to use Terraform to deploy a Pod to EKS:

```
resource "kubernetes_namespace" "example" {
  metadata {
    name = "example-namespace"
  }
}

resource "kubernetes_deployment" "example" {
metadata {
name = "example-deployment"
namespace = kubernetes_namespace.example.metadata.0.name
}

spec {
replicas = 2
```

```
template {
  metadata {
    labels = {
      app = "example"
    }
  }

  spec {
    container {
      name  = "example"
      image = "example-image:latest"
      port {
        name = "http"
        container_port = 80
      }
    }
  }
}
}
}

resource "kubernetes_service" "example" {
metadata {
name = "example-service"
namespace = kubernetes_namespace.example.metadata.0.name
}

spec {
selector = kubernetes_deployment.example.
spec.0.template.0.metadata.0.labels
port {
name = "http"
port = 80
target_port = "http"
}
}
}
```

This example creates a Kubernetes namespace, deployment, and service using Terraform. The deployment defines the container image, container name, container ports, and the number of replicas for the Pod. The service references the deployment and creates a load balancer that directs traffic to the Pods.

You can also use the `kubernetes_config_map` and `kubernetes_secret` resources to manage configuration data and secrets for the Pod.

It's worth noting that this is a simple example of deploying a Pod to EKS using Terraform. You can use the full set of options that the Kubernetes resources provide to create and manage more complex EKS environments, such as auto scaling, rolling updates, and integration with other AWS services such as CloudWatch, CloudTrail, and more.

You can also use the `terraform plan` and `terraform apply` commands to preview and apply the changes made to the EKS environment.

Summary

In conclusion, containers are a powerful tool for packaging and deploying applications in a consistent and portable way. AWS offers a variety of container services and platforms, each with its own set of features and capabilities. Terraform is an **infrastructure-as-code** (**IaC**) tool that can be used to manage and provision resources in AWS, including containers. By using Terraform to deploy containers to AWS, you can automate the process of creating and managing containerized applications, and ensure that your infrastructure is consistent, repeatable, and versionable. This can greatly simplify the process of deploying and scaling applications, and allows you to focus on the business logic of your application rather than managing the underlying infrastructure.

In the next chapter, we'll take a closer look at how Terraform can be leveraged for enterprise-level AWS projects. You'll learn about the unique challenges and considerations that come with managing large-scale infrastructure, and how to navigate the decision-making process when it comes to implementing AWS and Terraform at the enterprise level. We'll cover topics such as project planning, design considerations, and best practices for successful enterprise deployments. Stay tuned for a deep dive into the world of enterprise AWS and Terraform.

Part 3:
How to Structure and Advance Terraform in Enterprises

In this section, we explore how to use Terraform in enterprise-level projects, focusing on structuring and advancing Terraform implementations to meet the demands of large-scale organizations. We discuss how to integrate Terraform into enterprises, including building Git workflows for IaC and Terraform projects to enable version control, collaboration, and automated deployment. You'll learn how to automate the deployment of Terraform projects, streamlining the provisioning and management of cloud resources. We also delve into governance and security, exploring how to use Terraform to govern AWS resources and build a secure infrastructure on AWS. Finally, we discuss how to achieve a perfect AWS infrastructure with Terraform, optimizing performance, reliability, and cost-effectiveness. By the end of this part, you'll be equipped to structure and advance Terraform implementations in enterprises, ensuring scalable, secure, and efficient cloud infrastructure.

This part contains the following chapters:

- *Chapter 10, Leveraging Terraform for the Enterprise*
- *Chapter 11, Building Git Workflows for IaC and Terraform Projects*
- *Chapter 12, Automating the Deployment of Terraform Projects*
- *Chapter 13, Governing AWS with Terraform*
- *Chapter 14, Building a Secure Infrastructure with AWS Terraform*
- *Chapter 15, Perfecting AWS Infrastructure with Terraform*

10

Leveraging Terraform for the Enterprise

In the intricate world of enterprise-scale infrastructures, the balance between speed of deployment and operational efficiency is often a critical aspect. As organizations scale, the complexity of managing infrastructure, ensuring security compliance, and maintaining operational efficiency amplifies. Here, the advent of **infrastructure as code (IaC)** and cloud services have emerged as significant assets for system administrators, infrastructure engineers, and developers alike. Among the plethora of available tools and platforms, Terraform and **Amazon Web Services** (**AWS**) stand out for their versatility, reliability, and the robust ecosystem they offer.

In this insightful chapter, we delve deep into the intricate layers of employing Terraform in an AWS environment at an enterprise scale. We initiate our exploration by defining what an enterprise infrastructure project entails and how AWS magnifies its scope and potential. With AWS's vast and varied service offerings, comprehending its enterprise application can often seem overwhelming. Fear not, for we meticulously unfold this complex fabric to reveal a structured and manageable approach to navigating AWS enterprise projects.

Let's embark on this journey to unveil the synergies between Terraform and AWS, turning the complexities of enterprise-scale infrastructure management into a structured, efficient, and optimized endeavor.

In this chapter, we will cover the following topics:

- What is an enterprise infrastructure project?
- What is an AWS enterprise project?
- How to start an AWS enterprise project
- How to leverage Terraform in AWS enterprise projects
- How to decide/discuss/leverage AWS and Terraform implementations

What is an enterprise infrastructure project?

An enterprise infrastructure project refers to a large-scale, complex initiative aimed at upgrading, modernizing, or building the underlying technology systems and infrastructure of an organization. This can include hardware and software systems, data centers, networks, storage, and other critical components that support the day-to-day operations of the enterprise. The goal of an enterprise infrastructure project is to improve efficiency, reliability, and scalability while reducing costs and minimizing risk.

An enterprise infrastructure project is typically a multi-year, multi-disciplinary effort involving multiple teams and stakeholders. It may involve upgrading existing systems, replacing end-of-life components, or building new systems from scratch. The project will also involve planning, design, procurement, implementation, testing, and deployment phases, as well as ongoing support and maintenance activities. In many cases, the success of an enterprise infrastructure project depends on the ability to effectively manage risks, coordinate activities across teams, and ensure the timely delivery of high-quality solutions that meet business requirements. Additionally, the project should align with the organization's overall technology strategy and roadmaps and must consider factors such as security, compliance, and disaster recovery planning.

What is an AWS enterprise project?

An AWS enterprise project refers to a large-scale project that involves leveraging AWS to build, run, or manage the infrastructure or applications of an enterprise. This can include the migration of existing systems to the AWS cloud, the development of new applications, or the implementation of AWS-based solutions to support business requirements. The goal of an AWS enterprise project is to take advantage of the scalability, reliability, and cost-effectiveness of AWS to meet the needs of the business.

An AWS enterprise project can involve multiple services from the AWS portfolio, such as compute, storage, database, network, and security components, as well as integration with existing on-premises systems or other cloud providers. The project will also involve planning, design, deployment, testing, and ongoing management and optimization. Effective management of the AWS enterprise project requires expertise in cloud computing, AWS services, and enterprise IT architecture, as well as strong project management skills and the ability to manage risks and ensure that the project stays on track and within budget.

In addition to the technical aspects of the project, an AWS enterprise project also involves significant organizational and cultural changes, such as changes to existing processes, workflows, and roles. A successful AWS enterprise project requires strong leadership, communication, and collaboration among all stakeholders, including IT, business units, and executive leadership.

AWS provides a wide range of tools and services to support an AWS enterprise project, including management and security tools, development and deployment services, and a large ecosystem of partners and third-party solutions. This can help organizations to accelerate the project, minimize risks, and ensure the highest levels of security and compliance.

It's also important to consider ongoing costs and potential changes to the AWS bill, such as cost optimization strategies, to ensure that the benefits of the AWS enterprise project are sustainable in the long term. This can include leveraging automation, resource tagging, and cost allocation, as well as monitoring and reporting on usage and costs.

In summary, an AWS enterprise project is a complex and challenging undertaking that requires a combination of technical expertise, project management skills, and leadership to deliver business value while managing costs and risks.

How to define needs and solutions for an AWS enterprise project

Starting an AWS enterprise project can be a complex and challenging undertaking, but it can be simplified by following these steps:

1. **Define project goals and requirements**: Define the business goals and requirements that the project is expected to meet and ensure that these goals are aligned with the overall strategy and objectives of the enterprise.

2. **Assess existing systems and processes**: Assess the existing systems and processes in place, including any legacy systems that need to be migrated to the cloud.

3. **Develop a migration plan**: Develop a detailed migration plan that outlines the steps required to move existing systems and data to the AWS cloud. This plan should include a timeline, budget, and risk mitigation strategies.

4. **Choose AWS services**: Choose the AWS services that are most appropriate for the project, based on the requirements and goals defined in *step 1*.

5. **Design the AWS architecture**: Design the AWS architecture, including network, compute, storage, and security components, and ensure that the design meets the needs of the project.

6. **Implement security and compliance controls**: Implement the necessary security and compliance controls to ensure that the AWS environment is secure and compliant with industry and regulatory requirements.

7. **Test and deploy the infrastructure**: Test the AWS infrastructure, including any custom applications, and deploy to production once the testing is complete.

8. **Monitor and optimize**: Monitor the AWS infrastructure to ensure that it is operating as expected and continuously optimize to ensure that the infrastructure is cost-effective and meets the evolving needs of the business.

In conclusion, starting an AWS enterprise project requires careful planning, design, and implementation, as well as ongoing monitoring and optimization to ensure that the project is successful and provides the expected business value. It is important to have a team of experienced AWS professionals to manage the project, as well as strong leadership, communication, and collaboration among all stakeholders.

Here are some design considerations for an AWS enterprise project:

- **Requirements gathering**: Start by gathering requirements from all stakeholders, including business units, IT, and executive leadership. Identify the business objectives, technical requirements, and constraints for the project.

- **Architecture design**: Design a scalable, secure, and cost-effective AWS architecture that meets the requirements. Consider factors such as scalability, security, availability, performance, and cost-effectiveness when selecting AWS services and components.

- **Security and compliance**: Ensure that the AWS architecture meets all security and compliance requirements, including data privacy, security standards, and regulatory requirements. Consider using AWS security services such as Amazon GuardDuty, Amazon Inspector, and Amazon Certificate Manager to help with security.

- **Data migration**: Plan for the migration of data to AWS, including considerations for data transfer, data storage, and data backup and recovery. Use AWS services such as AWS Database Migration Service, AWS Snowball, and Amazon S3 to help with data migration.

- **Deployment automation**: Automate the deployment of the AWS architecture using tools such as AWS CloudFormation, AWS Elastic Beanstalk, and AWS CodeDeploy to minimize manual intervention and ensure consistent, repeatable deployments.

- **Cost optimization**: Optimize costs by selecting cost-effective services and pricing options, using tools such as AWS Cost Explorer, AWS Trusted Advisor, and AWS Reservations, and monitoring costs and usage regularly.

- **Monitoring and management**: Implement monitoring and management tools to ensure the availability, performance, and security of the AWS architecture. Use AWS services such as Amazon CloudWatch, AWS Systems Manager, and AWS Config to help with monitoring and management.

- **Training and support**: Provide training and support for IT staff and end users to ensure that they can effectively use and manage the AWS architecture.

In conclusion, starting an AWS enterprise project requires careful planning, design, and execution to ensure that the architecture is scalable, secure, cost-effective, and meets the requirements of the business. It is important to work with experienced AWS partners and consultants to help with the project, as well as to use AWS services and tools to support the project.

Defining success in AWS enterprise projects

Success in AWS enterprise projects depends on several factors, including careful planning, design, and execution, effective use of AWS services, and effective management and support. Here are some key steps for ensuring success in AWS enterprise projects:

1. **Define clear goals and objectives**: Start by defining clear goals and objectives for the project, including business objectives, technical requirements, and performance targets.

2. **Choose the right AWS services**: Choose the right AWS services to meet the requirements of the project. Consider factors such as scalability, security, availability, performance, and cost-effectiveness when selecting AWS services.

3. **Adopt a security-first approach**: Ensure that security and compliance are considered at all stages of the project, from design to deployment. Use AWS security services and tools to help with security and regularly assess and monitor security posture.

4. **Automate deployment and management**: Automate the deployment and management of the AWS architecture using AWS services and tools, such as AWS CloudFormation, AWS CodeDeploy, and AWS Systems Manager, to minimize manual intervention and ensure consistent, repeatable deployments.

5. **Monitor costs and optimize**: Monitor costs and optimize usage to minimize expenses and ensure cost-effectiveness. Use AWS cost optimization tools, such as AWS Cost Explorer, AWS Trusted Advisor, and AWS Reservations, to help with cost optimization.

6. **Provide training and support**: Provide training and support for IT staff and end users to ensure that they can effectively use and manage the AWS architecture. This includes training on AWS services, best practices, and troubleshooting techniques.

7. **Continuously monitor and improve**: Continuously monitor the AWS architecture and use feedback to identify areas for improvement. Regularly assess and update the architecture to ensure that it meets the evolving requirements of the business.

In conclusion, success in AWS enterprise projects requires a comprehensive approach that includes clear goals and objectives, the right AWS services, a security-first approach, automation, cost optimization, training and support, and continuous monitoring and improvement. It is important to work with experienced AWS partners and consultants to help ensure success.

How to discuss AWS enterprise projects

Discussing AWS enterprise projects requires clear communication, a deep understanding of the project goals and requirements, and the ability to effectively articulate the benefits and challenges of using AWS. Here are some tips for effectively discussing AWS enterprise projects:

- **Preparation**: Before discussing the project, research and understand the business requirements and objectives, as well as the AWS services that could be used to meet those requirements.

- **Start with the business requirements**: Start the discussion by highlighting the business requirements and how AWS can help meet those requirements. Emphasize the benefits of using AWS, such as scalability, security, and cost-effectiveness.

- **Explain the AWS services and architecture**: Explain the AWS services and architecture that will be used to meet the requirements, and highlight how the architecture is scalable, secure, and cost-effective. Use diagrams and visual aids to help explain the architecture.

- **Discuss the deployment and management approach**: Discuss the deployment and management approach, including how the architecture will be deployed and managed and how AWS services and tools, such as AWS CloudFormation, AWS CodeDeploy, and AWS Systems Manager, will be used to automate deployment and management.

- **Address any concerns or objections**: Address any concerns or objections raised by stakeholders, such as security, cost, or performance. Explain how AWS can help mitigate these concerns and provide references and case studies to support your argument.

- **Emphasize the benefits**: Emphasize the benefits of using AWS, including scalability, security, and cost-effectiveness, and explain how the benefits will help meet the business requirements and achieve the project goals.

In conclusion, discussing AWS enterprise projects requires effective communication and a deep understanding of the project goals, requirements, and benefits. By highlighting the benefits and addressing any concerns, you can effectively communicate the value of using AWS for enterprise projects.

How to leverage Terraform in AWS enterprise projects

Terraform is a popular open source IaC tool that can be leveraged in AWS enterprise projects to automate the provisioning, management, and versioning of cloud infrastructure. Here are some tips for leveraging Terraform in AWS enterprise projects:

- **Define the IaC**: Use Terraform to define the IaC, which enables you to version control and automate the provisioning and management of the infrastructure.

- **Use AWS provider**: Use the Terraform AWS provider to interact with AWS services and automate the deployment of the infrastructure. The AWS provider supports a wide range of AWS services, including EC2, RDS, S3, and VPC.

- **Automate the deployment process**: Automate the deployment process using Terraform, which enables you to deploy the infrastructure in a consistent and repeatable manner. This helps to reduce the risk of errors and increases the speed of deployment.

- **Use modules**: Use Terraform modules to encapsulate common infrastructure components and reduce duplication of code. Modules enable you to reuse infrastructure components across multiple projects, which can help to reduce time and effort.

- **Store Terraform state securely**: Store Terraform state securely in a centralized location, such as AWS S3 or AWS DynamoDB, to ensure that it can be accessed and updated by multiple teams and can be audited for compliance purposes.

- **Implement testing and validation**: Implement the testing and validation of Terraform code to ensure that it meets the requirements of the enterprise project. Use tools such as Terraform's built-in validation and testing capabilities, or other testing frameworks, to validate the infrastructure before it is deployed.

- **Collaborate with teams**: Collaborate with teams to ensure that Terraform is used consistently and effectively throughout the organization. Share best practices and collaborate on modules and templates to ensure that the infrastructure is deployed consistently and securely.

- **Integrate with other tools**: Integrate Terraform with other tools, such as configuration management tools (Ansible or Chef) or **continuous integration and continuous delivery (CI/CD)** tools, such as Jenkins or GitHub Actions, to automate the deployment pipeline and to ensure that changes are made in a consistent and controlled manner.

- **Consider multi-cloud deployments**: Consider multi-cloud deployments if the enterprise project requires a hybrid cloud approach. Terraform supports multiple cloud providers, including AWS, Azure, and **Google Cloud Platform (GCP)**, enabling you to automate the deployment and management of infrastructure across multiple clouds.

- **Plan for disaster recovery**: Plan for disaster recovery by using Terraform to automate the deployment of disaster recovery infrastructure, such as Amazon EC2 Auto Scaling groups and Amazon **Elastic File System (EFS)**, in addition to the primary infrastructure.

- **Security and compliance**: Ensure that Terraform code complies with enterprise security and compliance standards by using AWS services such as AWS **Key Management Service (KMS)** to store secrets and AWS Config to monitor the infrastructure.

- **Continuous improvement**: Continuously monitor and improve the Terraform infrastructure by tracking changes, monitoring the performance of the infrastructure, and continuously updating and refining the Terraform code.

In conclusion, leveraging Terraform in AWS enterprise projects enables you to automate the deployment and management of infrastructure, reduce the risk of errors, and ensure consistency and repeatability. By implementing testing and validation, storing the Terraform state securely, and collaborating with teams, you can ensure that Terraform is leveraged effectively in your AWS enterprise project.

Some recommendations for AWS enterprise projects

Here are some recommendations for using Terraform in AWS enterprise projects:

- **Store state securely**: Store the Terraform state securely, such as in AWS S3 or AWS DynamoDB, to ensure that it can be accessed and updated by multiple teams and can be audited for compliance purposes.

- **Implement testing and validation**: Implement testing and validation of Terraform code to ensure that it meets the requirements of the enterprise project. Use tools such as Terraform's built-in validation and testing capabilities, or other testing frameworks, to validate the infrastructure before it is deployed.

- **Collaborate with teams**: Collaborate with teams to ensure that Terraform is used consistently and effectively throughout the organization. Share best practices and collaborate on modules and templates to ensure that the infrastructure is deployed consistently and securely.

- **Use version control**: Use version control, such as Git, to manage Terraform code and to track changes to the infrastructure.

- **Plan for disaster recovery**: Plan for disaster recovery by using Terraform to automate the deployment of disaster recovery infrastructure, such as Amazon EC2 Auto Scaling groups and Amazon EFS, in addition to the primary infrastructure.

- **Security and compliance**: Ensure that Terraform code complies with enterprise security and compliance standards by using AWS services such as AWS KMS to store secrets and AWS Config to monitor the infrastructure.

- **Continuous improvement**: Continuously monitor and improve the Terraform infrastructure by tracking changes, monitoring the performance of the infrastructure, and continuously updating and refining the Terraform code.

In conclusion, using Terraform in AWS enterprise projects requires careful planning, collaboration with teams, and a focus on security and compliance. By storing the state securely, implementing testing and validation, and continuously monitoring and improving the infrastructure, you can ensure that Terraform is leveraged effectively in your AWS enterprise project.

Summary

In this chapter, we have unraveled the intricate landscape of employing Terraform in enterprise AWS environments, addressing strategic planning, security compliance, operational efficiency, and continuous improvement. Insights and strategies have been shared, enabling a transformation from complex enterprise-scale challenges to structured and optimized operations. As we transition to the next chapter, *Building Git Workflows for IaC and Terraform Projects*, we will delve into the integration of version control and collaborative development within IaC, ensuring consistency, traceability, and enhanced collaboration in managing and deploying infrastructure.

11
Building Git Workflows for IaC and Terraform Projects

In this chapter, we look at the pivotal role of Git workflows in managing **Infrastructure as Code (IaC)** and Terraform projects, particularly within **Amazon Web Services (AWS)** environments. We explore various Git workflows, offering insights to effectively implement them for optimized collaboration and code quality. This chapter provides a comprehensive guide on selecting, setting up, and managing Git workflows, along with tools specifically tailored for AWS and Terraform projects.

Security takes center stage as we share best practices to safeguard your Terraform projects, from backend security to **role-based access control (RBAC)** and compliance. We wrap up the chapter with strategic insights on streamlining AWS Terraform projects for enhanced efficiency and effectiveness.

In the subsequent chapters, we'll expand on advanced strategies and tools, empowering you to elevate the security, efficiency, and scalability of your IaC and Terraform projects in AWS environments to new heights.

In this chapter, we will cover the following topics:

- Why do we need a Git workflow?
- Implementing a Git workflow
- Tools and flows to use with AWS Terraform projects
- How to secure a Terraform project
- Streamlining AWS Terraform projects

Why do we need a Git workflow?

Git is a **version control system (VCS)** that allows multiple developers to work on the same code base while keeping track of changes and collaborating on code. A Git workflow is a set of guidelines that dictate how developers use Git to manage the code base.

There are several Git workflows, but the most common one is a **Gitflow** workflow. A Gitflow workflow is a branching model that provides a clear separation of development branches and release branches. It consists of two main branches:

- **Master branch**: The master branch represents the official code base, and it should always contain a stable, working version of the code.

- **Develop branch**: The develop branch is used for ongoing development work. Developers create feature branches off the develop branch, make changes to the code, and then merge their changes back into the develop branch.

In addition to the master and develop branches, there are also feature branches, release branches, and hotfix branches:

- **Feature branches** are used for developing new features. Developers create a new branch off the develop branch, make changes to the code, and then merge their changes back into the develop branch.

- **Release branches** are used for preparing a new release. Developers create a new branch off the develop branch, perform final testing and bug fixing, and then merge their changes into the master branch.

- **Hotfix branches** are used to fix critical issues in the master branch. Developers create a new branch off the master branch, make changes to the code, and then merge their changes back into the master and develop branches.

In addition to the Gitflow workflow, there are several other Git workflows that developers can use, depending on their needs and preferences. Here are some of the most common Git workflows:

- **Centralized workflow**: In a centralized workflow, all developers work on a single master branch, and changes are made directly to the master branch. While this workflow is simple and straightforward, it can lead to conflicts and make it difficult to manage changes in the code base.

- **Feature branch workflow**: In a feature branch workflow, developers create a new branch for each feature they work on. Once the feature is complete, they merge the changes back into the main branch. This workflow can be useful for teams that need to work on multiple features simultaneously.

- **Forking workflow**: In a forking workflow, each developer creates their own copy of the repository, known as a fork. They make changes to the code base in their fork and then submit a pull request to the main repository to merge their changes. This workflow is often used for open source projects where contributions come from many different developers.

No matter which workflow is used, Git provides a powerful set of tools for managing changes to the code base, collaborating with other developers, and ensuring that the code base remains stable and functional. By following a clear and consistent workflow, teams can work together more effectively and produce higher-quality code.

Overall, a Git workflow helps developers manage changes to the code base, collaborate effectively, and ensure that the code base is always stable and functional.

A Git workflow is important for several reasons:

- **Collaboration**: A Git workflow allows multiple developers to work on the same code base without interfering with each other's work. By following a consistent workflow, developers can manage changes to the code base in a clear and organized way, avoiding conflicts and making it easier to collaborate.

- **Change management**: A Git workflow allows developers to keep track of changes to the code base, including who made each change and when. This makes it easier to identify and resolve issues that arise during development and to roll back changes if necessary.

- **Code quality**: A Git workflow can help improve code quality by providing a clear process for code review and testing. By following a consistent workflow, developers can ensure that code changes are thoroughly reviewed and tested before being merged into the main branch.

- **Release management**: A Git workflow provides a clear process for preparing and releasing new versions of the code base. By using feature branches and release branches, developers can ensure that new features are properly tested and that the code base remains stable and functional during the release process.

- **Efficiency**: A Git workflow can help teams work more efficiently by reducing time spent resolving conflicts and managing changes to the code base. By following a clear and consistent workflow, developers can focus on writing code and collaborating with their team members, rather than managing the technical details of version control.

In summary, a Git workflow provides a set of guidelines that help teams manage changes to the code base, collaborate more effectively, and ensure that the code base remains stable and functional throughout the development process. By following a consistent workflow, teams can work more efficiently and produce higher-quality code.

Implementing a Git workflow

Implementing a Git workflow involves defining a set of guidelines and processes that dictate how developers will use Git to manage the code base. Here are general steps for implementing a Git workflow:

1. **Choose a Git workflow**: Choose the Git workflow that best suits the needs of your team and project. The most common Git workflow is Gitflow, but there are other workflows as well, such as a centralized workflow, a feature branch workflow, and a forking workflow.

2. **Set up the repository**: Create a Git repository to store the code base for your IAC project and set up the necessary branches, such as the master and develop branches.

3. **Define a process for creating and merging feature branches**: Define a process for creating and merging feature branches, such as naming conventions, coding standards, code review, and testing. Typically, developers will create a new branch off the develop branch for each feature they work on, make changes to the code, and then merge their changes back into the develop branch.

4. **Define a process for preparing and releasing new versions**: Define a process for preparing and releasing new versions of the code base. This may involve creating a release branch off the develop branch, performing final testing and bug fixing, and then merging changes into the master branch.

5. **Define a process for hotfixes**: Define a process for handling critical issues in the master branch. This may involve creating a hotfix branch off the master branch, making changes to the code, and then merging changes back into the master and develop branches.

6. **Train the team**: Train the development team on the Git workflow, including how to create and merge branches, how to handle conflicts, and how to use Git to manage the code base effectively.

7. **Review and iterate**: Continuously review and iterate on the Git workflow to ensure that it is working effectively and meeting the needs of the team and project.

Overall, implementing a Git workflow requires a clear understanding of the needs of the team and IaC project, as well as a willingness to experiment and iterate until the workflow is working effectively. With a well-defined Git workflow in place, teams can collaborate more effectively, manage changes to the IaC project code base more efficiently, and produce higher-quality templates.

Tools and flows to use with AWS Terraform projects

When using Terraform for AWS IAC projects, several Git tools can be used to implement a Git flow. Here are some of the most commonly used Git tools for Terraform:

- **Git**: Git is a popular VCS that can be used to manage changes to Terraform code. With Git, you can create branches for different features, manage changes to the code base, and collaborate with other developers.

- **GitHub**: GitHub is a popular Git hosting platform that provides features such as pull requests, code review, and collaboration tools. You can use GitHub to host your Terraform code and collaborate with other developers.

- **GitLab**: GitLab is another popular Git hosting platform that provides features such as **continuous integration/continuous delivery (CI/CD)** and security scanning. You can use GitLab to host your Terraform code, manage pipelines, and collaborate with other developers.

- **Bitbucket**: Bitbucket is a Git hosting platform that provides features such as pull requests, code review, and collaboration tools. You can use Bitbucket to host your Terraform code and collaborate with other developers.

When implementing a Git flow with Terraform, it's important to follow a consistent workflow that includes branching strategies, pull requests, and code review. This can help ensure that changes to the Terraform code base are properly tested and reviewed before being merged into the main branch. Additionally, you may want to consider using tools such as Terraform Cloud or AWS CodePipeline for managing infrastructure deployment and release management.

Here are some additional tips for implementing a Git flow with Terraform:

- **Branching strategy**: When creating a branching strategy for Terraform, consider using a Gitflow-like approach with branches such as feature, develop, release, and master. You may also want to create separate branches for different environments, such as staging and production.

- **Pull requests and code review**: Use pull requests and code review to ensure that changes to the Terraform code base are properly reviewed and tested before being merged into the main branch. This can help catch potential issues early and prevent them from impacting the infrastructure.

- **Automated testing**: Consider using automated testing tools such as Terratest or Kitchen-Terraform to automate testing of your Terraform code base. This can help catch potential issues before they make it to production.

- **Versioning**: Use versioning tools such as **Semantic Versioning (SemVer)** to manage versioning of your Terraform code base. This can help ensure that changes are properly tracked and that different versions of the infrastructure can be easily managed and maintained.

- **IaC best practices**: Follow best practices for IaC such as using modularity, separating configuration from code, and creating reusable modules. This can help ensure that your Terraform code base is scalable, maintainable, and secure.

By following these tips and using the right Git tools, you can implement a robust and effective Git flow for your Terraform AWS IAC projects. This can help ensure that your infrastructure is properly managed, versioned, and tested and that your team can work efficiently and collaboratively.

How to secure a Terraform project

Securing a Terraform project involves taking several steps to ensure that the infrastructure is properly configured and protected against security threats. Here are some best practices for securing a Terraform project:

- **Use a secure backend**: Terraform stores state information in a backend, which can be a remote service such as Amazon **Simple Storage Service (S3)** or Terraform Cloud. Make sure that the backend is properly secured, with appropriate access controls and encryption.

- **Use variables and secrets**: Use variables and secrets to store sensitive information such as API keys, passwords, and other secrets. Store these variables and secrets in a secure location such as AWS Secrets Manager or a secure configuration management tool.

- **Use secure networking**: Ensure that the network configuration for the infrastructure is properly secured, with appropriate firewalls, **network security groups (NSGs)**, and **virtual private networks (VPNs)** in place.

- **Follow least privilege**: Use the **principle of least privilege (PoLP)** to ensure that access to the infrastructure is properly controlled. Use RBAC to ensure that only authorized users have access to the infrastructure.

- **Monitor for changes**: Monitor the infrastructure for changes and unusual activity. Use automated tools such as AWS CloudTrail to track changes to the infrastructure and identify potential security threats.

- **Use security tools**: Use security tools such as vulnerability scanners and penetration testing tools to identify potential security threats and vulnerabilities in the infrastructure.

- **Regularly update and patch**: Regularly update and patch the infrastructure to ensure that it is protected against known security threats and vulnerabilities.

- **Use secure coding practices**: Use secure coding practices such as input validation, error checking, and output encoding to prevent security vulnerabilities such as injection attacks.

- **Enable auditing**: Enable auditing for your infrastructure to track changes and identify security threats. Use tools such as AWS Config to track changes to your infrastructure and monitor for security threats.

- **Use encryption**: Use encryption to protect sensitive data such as API keys, passwords, and other secrets. Use encryption tools such as AWS **Key Management Service (KMS)** to encrypt and store sensitive data.

- **Use multi-factor authentication (MFA)**: Use MFA to ensure that only authorized users have access to your infrastructure. Use MFA tools such as AWS MFA to add an extra layer of security to your infrastructure.

- **Implement disaster recovery (DR)**: Implement DR measures to ensure that your infrastructure can recover from security incidents and other disasters. Use tools such as AWS **Elastic Disaster Recovery (DRS)** to implement DR measures for your infrastructure.

- **Follow compliance standards**: Follow compliance standards such as the **Payment Card Industry Data Security Standard (PCI DSS)** or the **Health Insurance Portability and Accountability Act (HIPAA)** to ensure that your infrastructure meets the necessary security and compliance requirements.

By following these additional steps, you can further enhance the security of your Terraform project and help ensure that your infrastructure is properly protected against security threats. It's important to regularly review and update your security measures to ensure that your infrastructure remains secure over time.

Streamlining AWS Terraform projects

Streamlining an AWS Terraform project involves taking steps to optimize the infrastructure deployment process, reduce the time and effort required for deployment, and improve the efficiency of the development process. Here are some best practices for streamlining an AWS Terraform project:

- **Use modular code**: Use modular code to create reusable templates and modules that can be easily shared across the project. This can help reduce the amount of duplicated code and make it easier to maintain and update the infrastructure.

- **Use Terraform modules**: Use Terraform modules to encapsulate reusable infrastructure components such as security groups, load balancers, and databases. This can help simplify the infrastructure deployment process and reduce the time and effort required for deployment.

- **Use Terraform workspaces**: Use Terraform workspaces to manage multiple environments such as development, staging, and production. This can help streamline the deployment process and ensure that the infrastructure is properly configured for each environment.

- **Use Terraform Cloud**: Use Terraform Cloud to automate the deployment process and manage IaC. Terraform Cloud provides features such as collaboration, version control, and CI/CD, which can help streamline the development process.

- **Use automated testing**: Use automated testing tools such as Terratest or Kitchen-Terraform to automate the testing of your Terraform code base. This can help catch potential issues before they make it to production and reduce the amount of manual testing required.

- **Use CI/CD**: Use CI/CD tools such as Jenkins, GitLab CI/CD, or AWS CodePipeline to automate the deployment process and reduce the time and effort required for deployment.

- **Use IAC best practices**: Follow IAC best practices such as separating configuration from code, creating reusable modules, and using version control. This can help simplify the infrastructure deployment process and reduce the time and effort required for deployment.

- **Use tagging**: Use tagging to label and organize your infrastructure resources. This can help simplify management and make it easier to identify and manage resources.

- **Use AWS managed services**: Use AWS managed services such as Amazon **Relational Database Service** (**RDS**), Amazon ElastiCache, or Amazon **Simple Notification Service** (**SNS**) instead of managing infrastructure components manually. This can help simplify the deployment process and reduce the amount of manual work required.

- **Monitor infrastructure health**: Monitor the health of your infrastructure using Amazon CloudWatch. This can help identify potential issues early and prevent downtime.

- **Use parameterized templates**: Use parameterized templates to create reusable templates that can be customized for different environments. This can help simplify the deployment process and reduce the time and effort required for deployment.

By following these steps, you can streamline your AWS Terraform project and improve the efficiency of the development process. It's important to regularly review and update your infrastructure deployment process to ensure that it remains efficient and effective over time.

Summary

In this chapter, we unfolded the intricacies of integrating Git workflows into IaC and Terraform projects. We demystified the art of selecting and implementing robust Git workflows and elucidated security protocols essential for safeguarding your Terraform projects. We also navigated through strategies to enhance the efficiency of deploying AWS Terraform projects, setting the stage for advanced, streamlined, and secure infrastructure deployment.

As we transition into the next chapter, *Automating the Deployment of Terraform Projects*, prepare to delve deeper into the world of automation, where we'll explore cutting-edge tools and methodologies designed to optimize, expedite, and enhance the precision of deploying Terraform projects, turning complexity into simplicity and challenges into opportunities. We're on the brink of transforming theory into actionable insights!

12

Automating the Deployment of Terraform Projects

Automation and efficiency are key in today's fast-paced tech landscape. In this chapter, we'll be zeroing in on the automation of Terraform project deployments, elevating your **Infrastructure as Code (IaC)** practices to new heights.

We'll explore the core concepts of deployment in the Terraform context, shedding light on critical themes such as **continuous integration/continuous deployment (CI/CD)** in Terraform, and why it's an indispensable element of modern IaC practices. We'll unravel the complex web of choices to pinpoint the best CI/CD tools tailored for Terraform, guiding you through a sea of options so that you can find the one that aligns with your specific needs and organizational nuances.

We'll also venture into the intricate world of governance and auditability, offering you a roadmap to build systems that are not just efficient and automated but also secure, compliant, and audit-ready. Every piece of infrastructure provisioned will be a testament to best practices in security, efficiency, and compliance.

And as security is a paramount concern, this chapter doesn't shy away from the hard questions. We'll dive deep into strategies to ensure that every piece of infrastructure is provisioned securely, ensuring the sanctity and security of your organizational data and resources.

In essence, this chapter is your compass in the complex, multifaceted journey of automating Terraform deployments, a guide to transitioning from manual, error-prone processes to streamlined, efficient, and secure automated deployments.

In this chapter, we will cover the following topics:

- What is deployment in Terraform?
- What is CI/CD for Terraform?
- Why do we need CI/CD for Terraform?

- What is the best CI/CD tool for Terraform?
- How to build the governance and auditability of provisioning infrastructure
- How to provision infrastructure securely

What is deployment in Terraform?

In Terraform, deployment refers to the process of creating and configuring infrastructure resources using Terraform code. Terraform deployment involves creating and updating infrastructure resources such as virtual machines, databases, load balancers, and other resources.

The Terraform deployment process typically involves the following steps:

1. **Write Terraform code**: Write Terraform code that describes the desired infrastructure resources, including their configuration and dependencies.

2. **Plan the deployment**: Use the `terraform plan` command to create an execution plan that shows what changes Terraform will make to the infrastructure.

3. **Apply the changes**: Use the `terraform apply` command to apply the changes to the infrastructure. Terraform will create or update the infrastructure resources as necessary to match the desired state described in the Terraform code.

4. **Manage the infrastructure**: Once the infrastructure has been deployed, use Terraform to manage the infrastructure resources over time. This can include updating the configuration of existing resources, adding new resources, or deleting existing resources.

The Terraform deployment process can be automated using CI/CD tools. These tools can be used to manage the Terraform deployment process, including managing the Terraform code, creating execution plans, and deploying changes to the infrastructure.

By properly managing the Terraform deployment process, you can ensure that your infrastructure is properly configured, secure, and reliable. It's important to follow best practices for infrastructure deployment and to regularly review and update your deployment process to ensure that it remains efficient and effective over time.

What is CI/CD for Terraform?

CI/CD for Terraform involves using CI/CD tools to automate the deployment of infrastructure resources created using Terraform. CI/CD for Terraform is a process that involves the following stages:

1. **Continuous integration**: In the CI stage, changes made to the Terraform code base are automatically integrated into a shared repository. This can involve using version control tools such as Git to track changes to the Terraform code base and using automated testing tools to verify that the code changes are properly tested.

2. **Continuous delivery**: In the continuous delivery stage, changes made to the Terraform code base are automatically delivered to a test environment for further testing and verification. This can involve using tools such as AWS CodePipeline or GitLab CI/CD to automatically build and deploy the Terraform code base to a test environment.

3. **Continuous deployment**: In the continuous deployment stage, changes made to the Terraform code base are automatically deployed to the target production environment. This can involve using tools such as Terraform Cloud or AWS CodeDeploy to automate the deployment of infrastructure resources created using Terraform.

By using CI/CD for Terraform, you can automate the deployment of infrastructure resources and reduce the time and effort required for deployment. This can help ensure that the infrastructure is properly configured, secure, and reliable and that changes to the infrastructure are properly tested and reviewed before being deployed to the production environment.

Why do we need CI/CD tool for Terraform?

We need CI/CD for Terraform to automate the deployment process of infrastructure resources created using Terraform, and to ensure that changes to the infrastructure are properly tested and reviewed before being deployed to production. Here are some of the key benefits of using CI/CD for Terraform:

*   **Reduced time and effort**: By automating the deployment process regarding infrastructure resources, CI/CD can help reduce the time and effort required for deployment. This can help speed up the development process and reduce the risk of errors.

*   **Improved efficiency**: By automating the testing and deployment process, CI/CD can help improve the efficiency of the development process. This can help ensure that changes are properly tested and reviewed and that the infrastructure is properly configured and secured.

*   **Consistency and repeatability**: By using a consistent and repeatable process for deploying infrastructure resources, CI/CD can help ensure that the infrastructure is properly configured and that changes are properly tracked and managed over time.

*   **Improved collaboration**: By using CI/CD tools such as Terraform Cloud or AWS CodePipeline, developers can work collaboratively and share code and resources across the project. This can help improve the efficiency and effectiveness of the development process.

*   **Faster time to market**: By automating the deployment process, CI/CD can help speed up the time to market for new features and infrastructure resources. This can help organizations stay competitive and respond more quickly to changing business needs.

*   **Increased reliability**: By using CI/CD, organizations can increase the reliability of their infrastructure. With automated testing and deployment, developers can identify and fix issues quickly, reducing the likelihood of downtime or other issues.

- **Easier rollbacks**: With a consistent and repeatable deployment process, organizations can more easily roll back changes in the event of an issue or error. This can help reduce the impact of any issues and restore services more quickly.

- **Improved security**: CI/CD can help improve security by automating security testing and reviews as part of the deployment process. This can help identify potential security issues before they make it to production, reducing the risk of security incidents.

- **Reduced costs**: With CI/CD, organizations can reduce the costs associated with infrastructure deployment. By automating the deployment process, organizations can reduce the need for manual intervention and improve the efficiency of the development process.

- **Scalability**: With automated testing and deployment, organizations can more easily scale their infrastructure resources up or down to meet changing business needs. This can help organizations respond more quickly to changing demands and reduce the risk of downtime or other issues.

Overall, CI/CD can help organizations optimize their infrastructure deployment process and improve the efficiency and effectiveness of their development teams. By automating testing, deployment, and other key processes, organizations can reduce the time and effort required for deployment, improve the reliability and security of their infrastructure, and respond more quickly to changing business needs.

What is the best CI/CD for Terraform?

When selecting a CI/CD tool for Terraform, it's important to consider the specific needs and requirements of your organization. Here are some factors to consider when selecting a CI/CD tool for Terraform:

- **Integration with Terraform**: The CI/CD tool you choose should have strong integration with Terraform, allowing you to easily deploy infrastructure resources using Terraform code. It should be able to read and interpret Terraform configuration files and allow you to execute Terraform commands as part of the deployment process.

- **Compatibility with AWS**: If you're deploying infrastructure resources on AWS, you should choose a CI/CD tool that is compatible with AWS services and APIs. This will ensure that you can easily integrate your deployment process with other AWS services and take advantage of AWS-specific features and functionality.

- **Scalability**: Your CI/CD tool should be able to scale as your organization grows and your infrastructure requirements become more complex. This means it should be able to handle large-scale deployments, support parallelization, and provide other features that help streamline the deployment process.

- **Security**: Security should be a key consideration when selecting a CI/CD tool. Look for a tool that supports secure access and authentication, and that provides features such as encryption and audit trails to help keep your infrastructure secure.

- **Customizability**: Your CI/CD tool should be customizable to meet the specific needs of your organization. Look for a tool that allows you to configure and customize the deployment process, and that provides flexible deployment options such as rollbacks and incremental deployments.

Several CI/CD tools work well with Terraform on AWS, each with its strengths and capabilities. Here are some of the most popular CI/CD tools that are used for Terraform on AWS:

- **AWS CodePipeline**: AWS CodePipeline is a fully managed CI/CD service that supports Terraform, allowing you to easily automate the deployment of your infrastructure resources on AWS. CodePipeline integrates with a variety of AWS services, including CodeCommit, CodeBuild, and CodeDeploy, to provide a complete end-to-end solution.

- **Jenkins**: Jenkins is an open source automation server that supports Terraform and can be used to build, test, and deploy infrastructure resources on AWS. Jenkins has a large and active community, with many plugins available to extend its functionality.

- **GitLab CI/CD**: GitLab CI/CD is a complete DevOps platform that includes support for Terraform. GitLab CI/CD provides continuous integration, continuous delivery, and continuous deployment capabilities, making it a popular choice for teams that want an all-in-one solution.

- **CircleCI**: CircleCI is a cloud-based CI/CD service that supports Terraform, providing a scalable and flexible solution for automating the deployment of infrastructure resources on AWS. CircleCI supports parallelization, allowing you to speed up your build and deployment process.

- **Terraform Cloud** is a cloud-based, HashiCorp-built service that can be used as a CI/CD tool for Terraform. It provides a variety of features that can help automate the deployment of infrastructure resources and simplify the management of Terraform code.

You should find a tool that is well-suited to your organization's needs and requirements, and that can help you streamline the deployment process and improve the efficiency of your infrastructure management.

How to build the governance and auditability of provisioning infrastructure

Building the governance and auditability of provisioning infrastructure is important for several reasons. First, governance and auditability can help ensure that your infrastructure is compliant with regulatory requirements and industry best practices. This is critical for organizations that operate in regulated industries, where failure to comply with regulations can result in significant financial and reputational damage. By building governance and auditability into your infrastructure provisioning process, you can ensure that your infrastructure meets all the necessary regulatory requirements and is properly managed and audited.

Second, governance and auditability can help improve the security of your infrastructure. By enforcing **role-based access control** (**RBAC**) and code reviews, and implementing compliance checks and configuration drift detection, you can reduce the risk of unauthorized changes and potential security incidents. This can help protect your organization's sensitive data and resources and reduce the risk of data breaches and other security incidents.

Third, governance and auditability can improve the reliability of your infrastructure. By implementing testing, backup, and disaster recovery processes, you can ensure that your infrastructure is properly tested and validated before being deployed, and that data can be recovered in the event of an issue. This can help reduce the risk of downtime or other issues and improve the overall reliability and performance of your infrastructure.

Overall, building the governance and auditability of provisioning infrastructure is critical for organizations that rely on their infrastructure to operate effectively and securely. By implementing best practices and regular auditing and review, you can ensure that your infrastructure meets all the necessary compliance requirements, is properly secured, and is reliable and performant over time.

To build the governance and auditability of provisioning infrastructure with Terraform, you can follow these best practices:

- **Use version control**: Use version control systems such as Git to manage your Terraform code. This can help you track changes and maintain an audit trail of who made changes and when.

- **Enforce code reviews**: Enforce code reviews to ensure that changes are reviewed and approved by multiple people before being deployed. This can help catch potential issues and improve the quality of the infrastructure code.

- **Implement RBAC**: Implement RBAC to ensure that only authorized users have access to the Terraform code and deployment process. This can help improve security and reduce the risk of unauthorized changes.

- **Use a policy engine**: Use a policy engine such as **Open Policy Agent** (**OPA**) to enforce policies and best practices for your infrastructure. This can help ensure that your infrastructure is properly configured and secured.

- **Enable logging and monitoring**: Enable logging and monitoring for your infrastructure to track changes and detect potential issues. This can help you identify and remediate security incidents or other issues quickly.

- **Implement change management**: Implement change management processes to ensure that changes to the infrastructure are properly reviewed, approved, and documented. This can help ensure that changes are properly managed and that potential risks are identified and addressed.

- **Use a central Terraform repository**: Use a central Terraform repository to store your infrastructure code. This can help ensure that your code is properly managed and that changes are properly tracked and documented.

- **Implement configuration drift detection**: Implement configuration drift detection to detect when the actual infrastructure configuration does not match the desired state described in the Terraform code. This can help you identify and address configuration issues quickly.

- **Implement compliance checks**: Implement compliance checks to ensure that your infrastructure is compliant with relevant regulatory requirements and industry best practices. This can help reduce the risk of non-compliance and associated penalties.

- **Enable infrastructure testing**: Enable infrastructure testing to ensure that your infrastructure resources are properly tested and validated before being deployed. This can help reduce the risk of issues and improve the reliability of your infrastructure.

- **Implement backup and disaster recovery**: Implement backup and disaster recovery processes to ensure that your infrastructure is properly protected and that data can be recovered in the event of an issue. This can help ensure that your infrastructure is properly secured and reliable.

By following these best practices, you can build a robust governance and auditability framework for your Terraform infrastructure. It's important to regularly review and update your processes to ensure that they remain effective and aligned with your organization's needs and requirements.

How to provision infrastructure securely

Provisioning infrastructure securely with Terraform is important for several reasons. First, it helps to protect your organization's data and resources from unauthorized access and attacks. By implementing security controls, using secure communication protocols, and managing your credentials and secrets securely, you can help reduce the risk of data breaches and other security incidents.

Second, it can help ensure that your infrastructure is compliant with relevant regulations and industry best practices. This is particularly important for organizations that operate in regulated industries, where non-compliance can result in significant financial and reputational damage.

Finally, it can help improve the reliability and performance of your infrastructure. By using secure infrastructure as code principles, such as code reviews and version control, you can help ensure that your Terraform code is properly managed and that changes are reviewed and approved before being deployed. This can help reduce the risk of downtime or other issues and improve the overall reliability and performance of your infrastructure.

Overall, provisioning infrastructure securely with Terraform is critical for organizations that rely on their infrastructure to operate effectively and securely. By implementing best practices and regularly reviewing and updating your security practices, you can help ensure that your infrastructure is properly secured and compliant and that it performs reliably over time.

Provisioning infrastructure securely with Terraform involves several best practices and techniques. Here are some key steps to follow:

1. **Implement security controls**: Implement security controls such as firewalls, encryption, and access controls to help protect your infrastructure from unauthorized access and attacks. Use Terraform modules that have been designed with security in mind and use best practices such as limiting access to sensitive data and infrastructure resources.

2. **Securely store credentials**: Use a secure credential management solution to store and manage secrets and other sensitive information, such as API keys or passwords. Use encrypted storage and transport mechanisms to ensure that sensitive data is properly protected.

3. **Use secure communication protocols**: Use secure communication protocols such as HTTPS or SSH to communicate with your infrastructure resources. This can help ensure that your communications are properly secured and that your data is protected from eavesdropping or other attacks.

4. **Implement audit trails**: Implement audit trails to track changes to your infrastructure and monitor access to sensitive resources. This can help you identify potential security issues or compliance violations and take action to address them.

5. **Use secure IaC**: Use secure IaC principles to ensure that your Terraform code is secure and properly managed. This includes practices such as code reviews, version control, and RBAC to help ensure that your code is properly managed and that changes are reviewed and approved before being deployed.

6. **Implement regular vulnerability scanning**: Implement regular vulnerability scanning to identify potential security issues in your infrastructure and take action to address them. This can help reduce the risk of security incidents and ensure that your infrastructure is properly secured.

7. **Use secure Amazon Machine Images (AMIs)**: Use secure AMIs when deploying instances on AWS. Use images that are regularly updated with the latest security patches and that have been hardened according to industry best practices.

8. **Enable CloudTrail logging**: Enable CloudTrail logging to monitor AWS API activity and track changes to your infrastructure resources. This can help you identify potential security issues and audit changes to your infrastructure over time.

9. **Use virtual private clouds (VPCs) and security groups**: Use VPCs and security groups to help protect your infrastructure resources. Use security groups to restrict inbound and outbound traffic to your instances, and use VPCs to isolate your resources from the public internet.

10. **Implement secure data management**: Implement secure data management practices to ensure that your data is properly protected. Use encryption, access controls, and other techniques to help protect your data from unauthorized access and attacks.

11. **Use security-focused Terraform modules**: Use Terraform modules that have been designed with security in mind. Look for modules that implement security controls, such as encryption, access controls, and monitoring, to help ensure that your infrastructure is properly secured.

By following these best practices, you can help ensure that your infrastructure is properly secured and that potential security issues are identified and addressed quickly. It's important to regularly review and update your security practices to ensure that they remain effective and aligned with your organization's needs and requirements.

Summary

This chapter illuminated the intricate process of automating Terraform project deployments. You grasped essential concepts, including CI/CD for Terraform, effectively integrating it to enhance efficiency, security, and compliance in deploying infrastructure. This chapter delineated step-by-step processes, tools, and best practices, transforming the complex landscape of automation into an accessible, actionable roadmap.

Prepare to embark on a fascinating journey in the next chapter, where the power of Terraform meets the expansive, dynamic world of AWS. *Governing AWS with Terraform* unfolds the secrets of wielding Terraform's capabilities to manage, optimize, and govern AWS resources with precision, efficiency, and security.

Every AWS service, every resource, is about to become a playground where your mastery in Terraform shines, turning complexity into simplicity, and challenges into opportunities. Ready to transform your AWS management practices? The next chapter awaits, where every line of code is a step toward unparalleled governance, efficiency, and security on AWS. Stay tuned!

13

Governing AWS with Terraform

In this chapter, we will explore the concept of infrastructure governance and why it is crucial for managing AWS resources effectively. We will also dive into how Terraform can be used as a powerful tool for governing infrastructure. With the increasing complexity and scale of AWS projects, effective governance is essential to ensure security, compliance, cost-efficiency, and overall success. We will cover the fundamentals of infrastructure governance, the importance of AWS resource governance, tools for governing AWS Terraform projects, automation, and best practices for building cost-effective and secure AWS Terraform projects.

The following topics will be covered in this chapter:

- What is infrastructure governance?
- Why do we need infrastructure governance?
- How to govern infrastructure with Terraform
- How to use AWS tools with Terraform to govern IAC projects

What is infrastructure governance?

Infrastructure governance is the process of managing and controlling the use of IT resources, including hardware, software, and data. It is the practice of defining policies, procedures, and guidelines to ensure that IT resources are used efficiently, securely, and in compliance with regulatory requirements. In the context of cloud computing, infrastructure governance is the process of managing and controlling the use of cloud resources, such as servers, storage, and networking, to ensure that they are used effectively and efficiently.

The importance of infrastructure governance

Infrastructure governance is critical for organizations that want to ensure that their IT resources are used effectively and efficiently. Without effective governance, organizations may face several challenges, including the following:

- **Uncontrolled growth**: Without proper governance, organizations may end up with an uncontrolled and unmanageable IT environment, which can lead to inefficiencies, high costs, and security risks

- **Compliance issues**: In regulated industries, such as finance and healthcare, non-compliance with regulatory requirements can lead to severe penalties and damage to the organization's reputation

- **Security risks**: Without proper governance, organizations may not have adequate security measures in place to protect their IT resources from cyber threats

- **Lack of visibility**: Without proper governance, organizations may not have a clear understanding of their IT resources, which can make it difficult to make informed decisions and manage their IT environment effectively

Key elements of infrastructure governance

The key elements of infrastructure governance include policies, procedures, and guidelines to ensure that IT resources are used effectively and efficiently. Some of the essential elements of infrastructure governance are as follows:

- **Resource allocation**: Effective infrastructure governance requires allocating IT resources based on the organization's needs and priorities. This includes determining the appropriate level of resources required to support the organization's operations, as well as ensuring that resources are used efficiently and effectively.

- **Security**: Infrastructure governance must include policies and procedures to ensure that IT resources are protected from cyber threats, including data breaches, malware, and other types of attacks. This includes implementing appropriate security controls, such as firewalls, intrusion detection and prevention systems, and access controls.

- **Compliance**: Infrastructure governance must include policies and procedures to ensure that IT resources are used in compliance with regulatory requirements, industry standards, and best practices. This includes conducting regular audits and assessments to ensure that IT resources are compliant with applicable regulations.

- **Monitoring and reporting**: Effective infrastructure governance requires monitoring and reporting on the use of IT resources to ensure that they are being used effectively and efficiently. This includes tracking resource usage, identifying potential issues, and reporting on the status of IT resources to management.

Benefits of infrastructure governance

Effective infrastructure governance can provide several benefits to organizations:

- **Cost savings**: By ensuring that IT resources are used efficiently and effectively, infrastructure governance can help organizations save money on hardware, software, and other IT expenses

- **Improved security**: By implementing appropriate security controls, infrastructure governance can help organizations protect their IT resources from cyber threats

- **Compliance**: By ensuring that IT resources are used in compliance with regulatory requirements, infrastructure governance can help organizations avoid penalties and reputational damage

- **Improved decision-making**: By providing management with visibility into the organization's IT environment, infrastructure governance can help improve decision-making and enable more informed strategic planning

Overall, infrastructure governance is critical for organizations that want to effectively manage and control their IT resources. It helps ensure that IT resources are used efficiently, effectively, and securely, while also ensuring compliance with regulatory requirements and industry best practices. In the next few sections, we will discuss why infrastructure governance is essential for AWS resources and how Terraform can help organizations achieve effective infrastructure governance in AWS.

Why do we need infrastructure governance?

As organizations grow and infrastructure becomes more complex, it becomes increasingly difficult to manage and govern. Without proper governance, infrastructure can become unmanageable, leading to issues such as security vulnerabilities, compliance violations, and excessive costs. In this section, we will explore the importance of infrastructure governance and why it is crucial for modern organizations.

Governance is a critical aspect of managing any infrastructure, and this is especially true for cloud-based resources. AWS, with its vast array of services, provides a lot of flexibility and power for developers and operations teams, but it also requires careful governance to ensure resources are used efficiently, securely, and cost-effectively. Terraform provides a powerful tool for managing AWS infrastructure, but to truly govern AWS resources, it's important to understand the fundamentals of infrastructure governance, the importance of AWS resource governance, and the tools and automation available to govern AWS Terraform projects. This chapter will explore these topics in detail, providing you with a comprehensive understanding of how to build cost-effective, secure AWS Terraform projects that are governed according to best practices.

Security and compliance

Infrastructure governance helps organizations maintain security and compliance by ensuring that all resources are properly managed and secured. With proper governance, organizations can ensure that only authorized individuals have access to sensitive resources and that all resources are configured to meet regulatory requirements.

For example, if a company stores sensitive customer data in an AWS S3 bucket, it must ensure that the bucket is properly secured and that only authorized individuals have access. Without proper governance, the bucket could be misconfigured, leaving the data vulnerable to attack or theft.

Cost optimization

Infrastructure governance can also help organizations optimize costs by ensuring that resources are used efficiently and effectively. With proper governance, organizations can monitor resource usage and identify areas where resources can be optimized or eliminated.

For example, if a company has multiple AWS instances running, they may be able to consolidate those instances to save on costs. Without proper governance, it may be difficult to identify these cost-saving opportunities.

Standardization and consistency

Infrastructure governance helps organizations maintain standardization and consistency across their infrastructure. With proper governance, organizations can ensure that all resources are properly configured and follow the same set of standards and best practices.

For example, if a company has multiple AWS accounts, it can use Terraform to ensure that all accounts follow the same set of security and compliance policies. Without proper governance, it may be difficult to maintain consistency across multiple accounts.

Risk management

Infrastructure governance can also help organizations manage risk by identifying potential issues and taking proactive measures to mitigate those risks. With proper governance, organizations can monitor their infrastructure and identify potential security vulnerabilities or compliance violations before they become major issues.

For example, if a company is using AWS to store sensitive customer data, it can use Terraform to ensure that all resources are properly secured and meet regulatory requirements. Without proper governance, it may be difficult to identify potential risks and take proactive measures to mitigate them.

Infrastructure governance is essential for modern organizations to ensure security, compliance, cost optimization, standardization, consistency, and risk management. By implementing proper governance practices and using tools such as Terraform, organizations can maintain control over their infrastructure and avoid the many pitfalls that come with unmanaged infrastructure.

In this section, we explored the fundamentals of infrastructure governance, the importance of AWS resource governance, and the tools and automation techniques available for governing AWS Terraform projects. We learned that infrastructure governance is the set of policies, procedures, and practices used to manage and optimize the use of IT resources and that it is becoming increasingly important as organizations move toward cloud-based infrastructure. We also discussed the importance of AWS resource governance, which involves managing AWS resources to ensure compliance, cost optimization, and security.

In the next section, we will dive into how to govern infrastructure with Terraform. We will explore the features and benefits of Terraform and discuss how it can be used to implement infrastructure governance policies and procedures for AWS resources. We will also provide some best practices for using Terraform to govern infrastructure, including the use of modules, the adoption of a version control system, and the implementation of automated checks and peer reviews.

How to govern infrastructure with Terraform

Governance is a critical aspect of managing infrastructure at scale, and Terraform can be a powerful tool to help achieve it. Terraform provides a declarative way to manage **Infrastructure as Code (IaC)**, which makes it an ideal tool for infrastructure governance. This section will cover various best practices and strategies for governing AWS resources using Terraform.

To govern infrastructure with Terraform, it is crucial to establish a clear governance policy that defines the processes and procedures for managing infrastructure. This policy should include guidelines for resource creation, resource modification, resource deletion, resource versioning, and resource access control. It is also important to define roles and responsibilities for infrastructure management, including who is responsible for creating and modifying resources, who is responsible for approving changes, and who has access to sensitive resources.

Some of the critical areas to consider when governing infrastructure with Terraform include the following:

- **Resource provisioning**: Terraform provides a way to create, modify, and delete resources in a consistent and repeatable way. However, it is essential to establish guidelines for resource provisioning, including defining naming conventions, resource tagging, and resource categorization.

- **Resource versioning**: As infrastructure evolves, it is crucial to track changes to resources and maintain versioning history. Terraform enables the versioning of infrastructure code, which provides a clear audit trail of changes made to resources.

- **Resource access control**: Access control is essential to ensure that only authorized personnel can create, modify, or delete resources. Terraform integrates with AWS **Identity and Access Management (IAM)** to provide fine-grained access control to resources.

- **Compliance and security**: Compliance and security are critical considerations when governing infrastructure. Terraform provides various compliance and security features, including the ability to apply security policies to resources and scan infrastructure code for security vulnerabilities.

- **Automation**: Automation is critical to ensure consistent and repeatable infrastructure management. Terraform provides a way to automate infrastructure management tasks, including resource creation, resource modification, and resource deletion.

In the following sections, we will delve into each of these areas in more detail and provide guidance on how to govern AWS resources using Terraform.

Resource provisioning with Terraform

Resource provisioning is a fundamental aspect of infrastructure governance with Terraform. Terraform allows teams to define and provision resources in a declarative manner, which ensures that infrastructure remains consistent, secure, and cost-effective. By leveraging Terraform's resource provisioning capabilities, teams can automate the process of creating and updating infrastructure resources, which reduces the likelihood of human error and accelerates the deployment process.

Terraform's resource provisioning capabilities enable teams to define infrastructure resources using **HashiCorp Configuration Language (HCL)**, a **domain-specific language (DSL)** for defining IaC. HCL is easy to read and write, and it provides a high level of abstraction for defining infrastructure resources. This means that teams can focus on the business logic of their infrastructure without worrying about the underlying implementation details.

To provision resources with Terraform, teams typically follow these steps:

1. **Define resources**: The first step in resource provisioning is to define the resources that need to be provisioned. Terraform supports a wide range of resource types, including compute instances, databases, network components, and more. Teams define resources using Terraform's HCL syntax, which allows them to specify the resource type, properties, and dependencies.

2. **Plan changes**: After defining the resources, teams use Terraform to plan the changes that need to be made to the infrastructure. Terraform's planning functionality generates an execution plan that outlines the changes that will be made to the infrastructure resources. This plan can be reviewed and approved before the changes are applied, which provides an additional layer of governance.

3. **Apply changes**: Once the execution plan has been reviewed and approved, teams can apply the changes to the infrastructure. Terraform applies the changes safely and reliably, ensuring that resources are updated in the correct order and that errors are detected and handled gracefully.

Governance considerations for defining resources in Terraform

When defining resources in Terraform, it's essential to consider governance and compliance requirements. Here are some factors to keep in mind:

- **Resource naming conventions**: Establish naming conventions for resources to ensure consistency and avoid naming conflicts. Consider including a prefix that identifies the environment or projects the resource belongs to.

- **Resource tags**: Use tags to categorize and organize resources for cost allocation, resource management, and compliance purposes. Define tag policies that enforce standardization across the organization.

- **Resource types and configurations**: Choose resource types and configurations that comply with security and compliance requirements. For example, if you're deploying a database, ensure it's configured with appropriate security settings and access controls.

- **Approval workflows**: Establish workflows for approving resource deployments to ensure that changes are properly reviewed and authorized. Consider integrating Terraform with a change management system to track and manage changes to the infrastructure.

By considering these governance considerations, you can ensure that your Terraform infrastructure is deployed in a secure and compliant manner.

Managing access and permissions

One of the most important aspects of infrastructure governance is ensuring that access and permissions are managed correctly. Terraform provides several mechanisms for managing access and permissions to AWS resources. Let's take a look.

IAM roles and policies

Terraform provides a mechanism for defining IAM roles and policies within your IaC. By defining IAM roles and policies in Terraform, you can ensure that access to AWS resources is tightly controlled and that permissions are granted based on the principle of least privilege.

IAM roles can be created using the `aws_iam_role` resource type, while IAM policies can be created using the `aws_iam_policy` resource type. Once these resources are defined in Terraform, you can use them to grant permissions to specific users or groups within your organization.

AWS Organizations

If you have multiple AWS accounts within your organization, you can use AWS Organizations to manage access and permissions across all of your accounts. AWS Organizations provides a centralized way to manage policies, permissions, and billing across multiple accounts.

Terraform provides the `aws_organizations_account` resource type, which can be used to manage AWS accounts within an organization. You can use this resource to create and manage AWS accounts, as well as to define policies and permissions that apply across all of your accounts.

Cross-account access

If you need to grant access to resources across multiple AWS accounts, you can use cross-account access to do so. Cross-account access allows you to grant permissions to users or resources in one account to access resources in another account.

Terraform provides the `aws_iam_role` resource type, which can be used to define cross-account access. By defining a role in one account and granting permissions to that role, you can allow users or resources in another account to assume that role and access the resources that it has permission to access.

Resource-level permissions

In addition to managing access and permissions at the IAM and account level, it's also important to manage access and permissions at the resource level. Terraform provides several mechanisms for doing so:

- **Tags**: You can use tags to manage access and permissions to resources based on specific criteria, such as department or project.

- **VPC endpoints**: You can use VPC endpoints to manage access to AWS services from within your VPC. By defining VPC endpoints in Terraform, you can ensure that access to AWS services is controlled and that data doesn't leave your VPC.

- **Security groups**: You can use security groups to manage access to EC2 instances and other resources within your VPC. By defining security groups in Terraform, you can ensure that access to resources is tightly controlled and that permissions are granted based on the principle of least privilege.

Implementing security best practices

When managing infrastructure with Terraform, security should be a top priority. Here are some security best practices that can be implemented:

- **Use encryption**: Always encrypt sensitive data such as passwords, private keys, and API keys. Terraform allows you to use various encryption mechanisms such as AES and RSA to encrypt your sensitive data.

- **Limit access to sensitive data**: Restrict access to sensitive data such as AWS access keys and secret access keys. Avoid embedding AWS keys in plain text in your Terraform files. Instead, use a secure secrets management system such as AWS **Key Management Service** (**KMS**).

- **Secure communication**: Ensure that all communication between Terraform and your infrastructure is secure. This can be achieved by using SSL/TLS to encrypt your connections.

- **Secure remote state storage**: Always use secure storage for remote state data. Remote state data can be sensitive and should be protected. Terraform supports various storage backends, including Amazon S3, Google Cloud Storage, and Azure Blob Storage.

- **Enable logging and auditing**: Enable logging and auditing of all Terraform activities to track changes and identify security issues. Logging can be done through Terraform's logging capabilities or by integrating with third-party logging tools.

- **Use multi-factor authentication (MFA)**: Enable MFA for all users who access your Terraform infrastructure. MFA adds an extra layer of security by requiring a second factor, such as a mobile device or security token, in addition to a password.

- **Monitor your infrastructure**: Monitor your infrastructure regularly for security issues and vulnerabilities. Use Terraform's built-in monitoring capabilities or integrate with third-party monitoring tools to track changes and identify potential security issues.

By implementing these security best practices, you can ensure that your Terraform infrastructure is secure and protected against potential security threats.

Configuring logging and monitoring

Logging and monitoring are critical components of infrastructure governance. They help teams to track and troubleshoot issues, as well as to detect and respond to potential security breaches.

With Terraform, you can configure logging and monitoring for your AWS infrastructure in a centralized and automated way. You can use AWS CloudTrail to log AWS API calls and AWS Config to monitor compliance with your desired configurations. You can also integrate with third-party logging and monitoring tools, such as Datadog or Splunk, to get more advanced insights and alerts.

To configure logging and monitoring with Terraform, you need to define the necessary resources in your configuration. For example, to enable CloudTrail, you can use the `aws_cloudtrail` resource and specify the S3 bucket where the logs should be stored. Similarly, to enable AWS Config, you can use the `aws_config_configuration_recorder` resource and specify the rules and resources to monitor.

It's also important to ensure that your logs and monitoring data are secured and encrypted. You can use AWS KMS to manage your encryption keys and encrypt your data at rest and in transit. You can also define IAM roles and policies to control access to your logs and monitoring data.

Overall, logging and monitoring are critical for infrastructure governance and should be an integral part of your Terraform configuration. By defining these resources in code, you can ensure that they are consistent, scalable, and automated across your AWS infrastructure.

Establishing resource naming conventions

Resource naming conventions are important for tracking and identifying resources in your infrastructure. Naming conventions must be clear and consistent to make it easier to identify resources, prevent naming conflicts, and support automation.

Here are some best practices for establishing resource naming conventions in Terraform:

- Use a standardized naming convention that is easy to read and understand, such as `{Environment}-{ResourceType}-{Name}`

- Keep resource names short but descriptive, and use only lowercase letters, numbers, and hyphens

- Use consistent and meaningful names for similar resources, such as `"web-server-1"` and `"web-server-2"` for two web servers in a cluster

- Use logical grouping to separate resources based on their function, such as `"network-"` for network-related resources and `"compute-"` for compute-related resources

- Use unique identifiers for resources that have the same or similar names, such as `"db-instance-1"` and `"db-instance-2"` for two database instances

- Use variables to enable dynamic naming of resources, such as prefixing the resource name with the environment name or project name

By following these resource naming conventions, you can make it easier to identify, manage, and monitor resources in your infrastructure.

Using version control and collaboration tools

Infrastructure governance with Terraform is a collaborative effort, and version control tools play a critical role in managing changes. Teams can use version control tools to track changes, collaborate, and manage the development and deployment of IaC. Here are some tips to effectively use version control and collaboration tools for your Terraform projects:

- **Use Git for version control**: Git is one of the most widely used version control tools available. It is easy to use and integrates well with most other tools in the DevOps and infrastructure management ecosystem.

- **Create a centralized Git repository**: A centralized Git repository makes it easy to manage changes across the team. All team members can access the same repository, review changes, and make updates as needed.

- **Use branching**: Branching allows teams to work on separate versions of the infrastructure simultaneously. This helps minimize conflicts and ensure that changes are reviewed before they are merged into the main branch.

- **Implement a code review process**: Code reviews are an essential part of the collaboration process. Code reviews help ensure that changes are properly reviewed and tested before they are merged into the main branch.

- **Use automation tools to enforce policies**: Automation tools such as Checkov or Sentinel can be used to enforce policies, scan code for vulnerabilities, and ensure that infrastructure code adheres to best practices.

- **Establish collaboration practices**: Teams should establish collaboration practices that define how code is reviewed, tested, and merged. This helps ensure that everyone is working together consistently and productively.

- **Use communication tools**: Communication tools such as Slack or Microsoft Teams can be used to keep everyone on the team informed about changes, issues, and other important information related to the Terraform project.

By following these tips, teams can effectively manage changes to their Terraform infrastructure code, collaborate, and ensure that best practices are followed.

Building and deploying with automation and pipelines

Automating the build and deployment process for IaC projects is an essential part of governance. Automation ensures that the build and deployment processes are predictable and repeatable, reducing the risk of human error and increasing the speed of development.

Pipelines are the foundation of automation in IaC projects. A pipeline is a series of steps that are executed in sequence to build, test, and deploy the infrastructure. Pipelines typically include stages for linting, testing, building, and deploying the infrastructure. The steps in each stage are executed in sequence, and if any step fails, the entire pipeline is aborted.

To implement a pipeline for your IaC project, you will need to choose a pipeline tool that integrates with your version control system and infrastructure platform. Some popular pipeline tools for IaC projects include Jenkins, GitHub Actions, Terraform Cloud, Terraform Enterprise, and GitLab CI/CD.

Once you have chosen a pipeline tool, you will need to define the stages and steps in your pipeline. Each step in the pipeline should be defined as a separate script or executable that can be run independently. This makes it easy to test and debug individual steps and also makes it easier to maintain and update the pipeline as your infrastructure changes.

To ensure that your pipeline is secure, you should use secrets management tools to store and manage your credentials and other sensitive information. You should also use automated testing tools to ensure that your infrastructure is secure and compliant with your organization's policies and standards.

Overall, building and deploying IaC with automation and pipelines is a critical part of governance. Automation ensures that the build and deployment processes are predictable and repeatable, while pipelines provide a framework for testing, building, and deploying the infrastructure. By implementing automation and pipelines in your IaC projects, you can reduce the risk of human error, increase the speed of development, and ensure that your infrastructure is secure and compliant with your organization's policies and standards.

Tracking and managing costs and budgets

Tracking and managing costs and budgets is an important part of infrastructure governance. Terraform provides several features that can help manage costs and track expenses.

One way to track costs is to use Terraform's ability to set budgets and configure alerts based on cost metrics. With AWS, Terraform can integrate with the AWS Budgets service to set and track budgets, as well as send notifications when budgets are exceeded.

Another way to manage costs is to use Terraform's ability to provision infrastructure based on specific cost requirements. For example, using the `aws_instance` resource, it's possible to specify the `instance_type` parameter to provision instances that fit within specific price ranges.

In addition to Terraform's built-in cost management features, there are also third-party tools that can help manage costs and expenses. CloudHealth by VMware and CloudCheckr are two popular options that integrate with Terraform and provide additional cost management and optimization features.

Overall, by implementing cost management practices with Terraform, organizations can ensure that they are using their resources efficiently and staying within budget.

Implementing compliance and governance policies

In addition to security, compliance and governance policies are crucial to ensuring the proper functioning and management of your infrastructure. Terraform provides numerous tools and features to help you ensure compliance with various regulations and standards, such as HIPAA, PCI DSS, and SOC 2.

To implement compliance and governance policies with Terraform, you can use the following tools:

- **Sentinel**: Sentinel is a policy-as-code framework built into Terraform Enterprise. It enables you to define and enforce policies across all of your IaC, using a familiar programming language.

- **Open Policy Agent (OPA)**: OPA is a flexible and lightweight policy engine that can be used to enforce policies across your IaC. OPA is compatible with Terraform and can be used to define policies for Terraform configurations and plans.

- **AWS Config**: AWS Config is a service that enables you to assess, audit, and evaluate the configurations of your AWS resources. You can use AWS Config to monitor compliance with regulatory standards and best practices and to enforce governance policies across your AWS infrastructure.

By implementing compliance and governance policies with Terraform, you can ensure that your infrastructure is secure, reliable, and compliant with regulatory standards and best practices.

In conclusion, infrastructure governance is an essential aspect of managing cloud resources, especially when dealing with large and complex environments. Terraform provides a powerful platform to implement governance policies and automate infrastructure management, allowing organizations to achieve cost savings, security, and compliance goals. By following best practices for resource provisioning, access and permissions, security, logging and monitoring, resource naming, version control and collaboration, automation and pipelines, cost tracking, and compliance policies, organizations can establish a strong governance framework for their AWS infrastructure. With the right tools and processes in place, teams can ensure that their AWS Terraform projects are secure, efficient, and cost-effective.

Summary

In this chapter, we explored the importance of infrastructure governance and how it can be achieved with Terraform on AWS. First, we defined what infrastructure governance is and why it is essential to have proper governance policies in place. Then, we explored how Terraform can be used to govern infrastructure by defining resources, managing access and permissions, implementing security best practices, configuring logging and monitoring, establishing resource naming conventions, using version control and collaboration tools, building and deploying with automation and pipelines, and tracking and managing costs and budgets.

We also discussed how to implement compliance and governance policies to ensure that infrastructure is managed in a compliant and secure manner. By following these best practices, organizations can build cost-effective, secure, and compliant AWS infrastructure while leveraging the benefits of Terraform.

In conclusion, Terraform provides a powerful toolset for governing infrastructure on AWS, and by following the best practices outlined in this chapter, organizations can maintain a high level of security, compliance, and efficiency in their infrastructure management.

As this chapter concludes, we turn our attention to the next challenge: building a secure infrastructure with Terraform, laying the groundwork for a resilient and scalable digital environment.

14

Building a Secure Infrastructure with AWS Terraform

In today's fast-paced and dynamic world, where technology is evolving rapidly, securing infrastructure has become a top priority for organizations. With the increase in cyber threats and attacks, building a secure infrastructure is crucial for protecting sensitive data and ensuring business continuity.

If you're looking to build a secure infrastructure on AWS, Terraform is an excellent choice. Terraform provides a platform-agnostic and declarative approach to **infrastructure as code (IaC)** that simplifies the process of building and managing secure infrastructure.

In this chapter, we'll discuss the importance of security in infrastructure, the best practices for governing security in AWS, and how to build a secure infrastructure with Terraform. We'll also explore the relationship between security and Terraform, as well as the benefits of using Terraform for building secure infrastructure.

By the end of this chapter, you'll have a solid understanding of the security fundamentals for cloud and AWS, the skills to govern security in AWS with Terraform, and the ability to build a secure infrastructure with Terraform. You'll also learn about audit trails and how to secure infrastructure operations and provisions.

We will cover the following topics:

- What is security in infrastructure?
- How to govern security in AWS
- How to build secure infrastructure in Terraform
- Security and Terraform
- Security and IaC operations

Let's dive into the world of securing infrastructure with AWS Terraform and take your security game to the next level.

What is security in infrastructure?

Security is one of the most important considerations when building any infrastructure. In the context of IT infrastructure, security refers to the measures and techniques that are put in place to protect the infrastructure and the data it holds from unauthorized access, theft, destruction, and other malicious activities. Building a secure infrastructure is essential for any organization, especially those that deal with sensitive information, such as financial or medical data.

In this section, we'll discuss the various aspects of security in infrastructure and what it entails.

By the end of this section, you should have a clear understanding of what security in infrastructure means and what measures are necessary to build a secure infrastructure on AWS using Terraform.

Threats to infrastructure security

IT infrastructure is vulnerable to a range of threats, both external and internal. These threats can compromise the integrity, confidentiality, and availability of infrastructure resources and the data they hold. Here are some common threats to infrastructure security:

- **Malware and viruses**: Malware and viruses are malicious software programs designed to infiltrate systems and steal sensitive data, disrupt operations, or damage hardware

- **Phishing and social engineering**: Phishing and social engineering attacks aim to trick users into providing sensitive information, such as login credentials, by impersonating legitimate entities or using other deceptive tactics

- **Unauthorized access**: Unauthorized access occurs when an attacker gains access to the infrastructure and data without proper authorization, either by exploiting vulnerabilities or using stolen credentials

- **Insider threats**: Insider threats are malicious or unintentional actions taken by authorized users, such as employees or contractors, that compromise the security of the infrastructure and data

- **Denial-of-service (DoS) attacks**: DoS attacks overload the infrastructure with traffic or requests, making it unavailable to legitimate users

- **Ransomware**: Ransomware is a type of malware that encrypts data and demands a ransom payment in exchange for the decryption key

- **Advanced persistent threats (APTs)**: APTs are sophisticated and targeted attacks that are designed to gain access to infrastructure and data over an extended period, often by using multiple attack vectors

- **Zero-day vulnerabilities**: These are software flaws that are unknown to the vendor and can be exploited by attackers before a patch or update is released

- **Data breaches**: Data breaches occur when sensitive data is accessed, stolen, or disclosed without authorization

- **Physical threats**: Physical threats to infrastructure security include theft, damage, or destruction of hardware, such as servers or networking equipment

It's essential to recognize and understand the various threats to infrastructure security to develop an effective security strategy. A comprehensive security strategy should address all possible threats and vulnerabilities and implement appropriate security measures to mitigate the risks.

In the following section, we will explore how to use AWS and Terraform to build a secure infrastructure that can protect against these threats and implement best practices for securing infrastructure.

The importance of infrastructure security

Infrastructure security is crucial for maintaining the integrity, confidentiality, and availability of data and resources. Without proper security measures in place, infrastructure is vulnerable to attacks and data breaches that can have severe consequences for organizations, such as the following:

- **Financial loss**: Data breaches and other security incidents can result in significant financial losses for organizations, including the cost of remediation, regulatory fines, and legal fees

- **Reputational damage**: Security incidents can also damage an organization's reputation, erode customer trust, and result in lost business

- **Legal and compliance issues**: Organizations that fail to protect sensitive data can face legal and regulatory consequences, including fines, lawsuits, and damage to their brand and reputation

- **Disruption of operations**: Security incidents can also disrupt business operations, resulting in lost productivity, revenue, and customer satisfaction

Given the potential impact of security incidents, organizations need to prioritize infrastructure security and implement best practices for securing infrastructure. In the next section, we'll discuss some of the basic principles of infrastructure security that can help organizations protect against threats and ensure the integrity, confidentiality, and availability of data and resources.

Basic principles of infrastructure security

To ensure the security of infrastructure, it's important to follow some basic principles of infrastructure security:

- **Defense in depth**: Defense in depth is a strategy that involves implementing multiple layers of security to protect infrastructure and data. This approach can help organizations reduce the risk of security incidents and limit the impact of any incidents that do occur.

- **Least privilege**: Least privilege is a security principle that involves giving users and processes only the minimum access required to perform their tasks. This principle can help organizations limit the impact of security incidents and prevent unauthorized access to infrastructure and data.

- **Encryption**: Encryption is the process of encoding data so that it can only be read by authorized parties. This principle can help organizations protect sensitive data, even if it is accessed by unauthorized parties.

- **Monitoring and logging**: Monitoring and logging are essential for detecting and responding to security incidents. Organizations should implement robust monitoring and logging solutions to track user and system activity and identify potential security incidents.

- **Continuous improvement**: Security is an ongoing process, and organizations should continually evaluate and improve their security posture. This includes regularly updating security measures and protocols, performing security assessments and audits, and staying up to date with the latest security trends and best practices.

By following these basic principles of infrastructure security, organizations can reduce the risk of security incidents, protect sensitive data, and ensure the integrity, confidentiality, and availability of infrastructure resources.

Types of security measures for infrastructure

To protect against threats to infrastructure security, organizations should implement various security measures and protocols. Here are some of the most common types of security measures for infrastructure:

- **Access control**: Access control measures help organizations limit access to infrastructure and data to authorized users only. These measures can include **multi-factor authentication** (**MFA**), **role-based access control** (**RBAC**), and network segmentation.

- **Firewall and network security**: Firewall and network security measures help organizations protect against unauthorized access to infrastructure and data by filtering traffic and enforcing security policies.

- **Antivirus and malware protection**: Antivirus and malware protection measures help organizations detect and remove malicious software from infrastructure and data.

- **Data backup and recovery**: Data backup and recovery measures help organizations protect against data loss due to security incidents, hardware failures, or other issues.

- **Patch and vulnerability management**: Patch and vulnerability management measures help organizations ensure that infrastructure and software are up to date and free of known vulnerabilities that can be exploited by attackers.

- **Incident response**: Incident response measures help organizations detect, contain, and respond to security incidents promptly and effectively.

By implementing these types of security measures for infrastructure, organizations can significantly reduce the risk of security incidents and ensure the integrity, confidentiality, and availability of infrastructure resources and data. In the next few sections, we'll explore the relationship between security and Terraform and how Terraform can help organizations build a secure infrastructure on AWS.

The role of governance in infrastructure security

Governance is a critical aspect of infrastructure security. Governance involves the policies, procedures, and processes that an organization puts in place to ensure that infrastructure resources are used effectively, efficiently, and securely. Here are some key ways that governance can help organizations improve infrastructure security:

- **Standards and policies**: Governance frameworks can provide standards and policies for securing infrastructure resources and data. These standards and policies can help ensure that infrastructure resources are configured securely and that security protocols are followed consistently.

- **Risk management**: Governance frameworks can help organizations identify and manage risks to infrastructure security, including vulnerabilities, threats, and compliance issues.

- **Compliance**: Governance frameworks can help organizations comply with relevant laws, regulations, and standards related to infrastructure security and data privacy.

- **Training and awareness**: Governance frameworks can provide training and awareness programs to help employees and stakeholders understand the importance of infrastructure security and the role they play in maintaining it.

By implementing robust governance frameworks, organizations can ensure that security is integrated into every aspect of infrastructure management. This can help organizations build a culture of security and ensure that security is a top priority at all times. In the next section, we'll explore how to govern security in AWS.

How to govern security in AWS

Now that we've explored the basics of infrastructure security and the role of governance in securing infrastructure resources, let's turn our attention to how to govern security in AWS. AWS provides a range of security features and services to help organizations build and manage secure infrastructure. However, to ensure that security is integrated into every aspect of AWS management, organizations should also implement robust governance frameworks that align with their security objectives.

By the end of this section, you should have a solid understanding of how to govern security in AWS.

AWS security services and features

AWS provides a range of security services and features that can help organizations build and manage secure infrastructure on the cloud. Let's have a look at some of these services and features:

- **AWS Identity and Access Management (IAM)**: IAM is a service that enables organizations to manage access to AWS resources securely. With IAM, organizations can create and manage user accounts, roles, and groups, and control permissions to access AWS resources.

- **AWS Key Management Service (KMS)**: KMS is a managed service that makes it easy to create and control the encryption keys that are used to encrypt data stored in AWS services and customer applications.

- **AWS Certificate Manager (ACM)**: ACM is a service that provides SSL/TLS certificates for use with AWS services and applications. With ACM, organizations can easily provision, manage, and deploy SSL/TLS certificates for their infrastructure.

- **AWS Firewall Manager**: Firewall Manager is a service that enables organizations to centrally manage and configure AWS **Web Application Firewall** (**WAF**) rules across multiple accounts and resources.

- **AWS GuardDuty**: GuardDuty is a threat detection service that continuously monitors malicious activity and unauthorized behavior in AWS accounts and workloads.

- **AWS Security Hub**: AWS Security Hub is a security service that provides a comprehensive view of security alerts and compliance status across an organization's AWS accounts. With Security Hub, organizations can aggregate and prioritize security findings from various AWS services, such as AWS GuardDuty, AWS Inspector, and Amazon Macie. Security Hub also provides automated compliance checks against industry standards, such as CIS AWS Foundations Benchmark and **Payment Card Industry Data Security Standard** (**PCI-DSS**).

These are just a few examples of the security services and features provided by AWS. By leveraging these services and features, organizations can improve the security of their AWS infrastructure and ensure that their data and resources are protected from unauthorized access and attacks.

AWS security compliance and certifications

AWS adheres to various security compliance standards and certifications to ensure that customer data and infrastructure are protected against security threats. Let's look at some of these compliance standards and certifications:

- **Service Organization Control 2 (SOC 2)**: This is an auditing procedure that verifies that AWS has appropriate controls and procedures in place to protect customer data and infrastructure against security threats.

- **Health Insurance Portability and Accountability Act (HIPAA)**: This is a US law that sets standards for the security and privacy of electronic health information. AWS provides services that can help customers comply with HIPAA requirements.

- **PCI-DSS**: This is a set of security standards that govern the processing, storage, and transmission of credit card information. AWS provides services that can help customers comply with PCI DSS requirements.

- **ISO/IEC 27001**: ISO/IEC 27001 is a widely recognized international standard for information security management. AWS has been certified as ISO/IEC 27001 compliant, demonstrating its commitment to maintaining robust security practices and procedures.

By adhering to these compliance standards and certifications, AWS can provide customers with a secure and reliable platform for their infrastructure and data. Additionally, customers can leverage these compliance standards and certifications to demonstrate their compliance with relevant laws and regulations.

AWS maintains compliance with various security compliance standards and certifications by conducting regular audits, assessments, and evaluations. AWS undergoes independent third-party audits that evaluate its controls against security and compliance standards.

Additionally, AWS performs internal assessments to evaluate and improve its security posture, including its policies, procedures, and controls. AWS also provides various tools and services that can help customers achieve and maintain compliance with these security and compliance standards. These tools and services include AWS Artifact, which provides customers with on-demand access to AWS compliance reports and other compliance-related documents, and AWS Control Tower, which provides customers with a pre-configured environment that conforms to best practices for security and compliance. By maintaining compliance with these standards and certifications, AWS can provide customers with a secure and reliable platform for their infrastructure and data.

AWS security governance frameworks

AWS provides various governance frameworks and best practices that organizations can use to govern security in their AWS infrastructure. These frameworks and best practices include the following:

- **AWS Well-Architected Framework**: The AWS Well-Architected Framework provides a set of best practices for designing and operating reliable, secure, efficient, and cost-effective systems in the cloud. The framework includes a Security pillar that provides guidance on how to implement security best practices in AWS infrastructure.

- **AWS Security Hub**: As mentioned earlier, AWS Security Hub provides a comprehensive view of security alerts and compliance status across an organization's AWS accounts. With Security Hub, organizations can centrally manage compliance checks and automate the response to security incidents, making it easier to identify and remediate security issues in AWS infrastructure.

- **AWS Control Tower**: This is a service that provides a pre-configured environment that conforms to best practices for security and compliance. Control Tower automates the setup and management of multiple AWS accounts, providing a centralized view of infrastructure and security compliance across an organization's AWS environment.

By implementing these AWS security governance frameworks and best practices, organizations can ensure that their AWS infrastructure is designed and operated securely and that security is integrated into every aspect of their AWS management.

Monitoring and logging for AWS security

Monitoring and logging are critical components of an effective security strategy in AWS. By monitoring and logging AWS infrastructure and services, organizations can detect and respond to security incidents promptly and identify trends and patterns in security-related events that can help improve overall security posture. Here are some AWS tools and services that can be used for monitoring and logging in AWS:

- **Amazon CloudWatch**: CloudWatch is a monitoring and observability service for AWS resources and applications. With CloudWatch, organizations can monitor metrics, collect and store log files, and create alarms to alert when certain conditions are met.

- **AWS Config**: AWS Config is a service that provides a detailed inventory of resources in an AWS account and also tracks changes to these resources over time. With Config, organizations can monitor the configuration of their infrastructure and ensure that it adheres to best practices for security and compliance.

- **AWS CloudTrail**: CloudTrail is a service that provides a record of events and activity within an AWS account. CloudTrail logs events such as API calls, AWS Management Console sign-ins, and AWS service events and can be used to detect unauthorized access and other security incidents.

By using these monitoring and logging tools and services, organizations can gain valuable insights into their AWS infrastructure and services and also improve their security posture by detecting and responding to security incidents promptly.

Incident response for AWS security

Despite the best efforts to secure AWS infrastructure, security incidents can still occur. Therefore, it's important to have an effective incident response plan in place to detect, respond to, and recover from security incidents in AWS. Here are some best practices for incident response in AWS:

- **Develop an incident response plan**: Develop a comprehensive incident response plan that outlines the steps to be taken in the event of a security incident. The plan should include roles and responsibilities, communication protocols, and escalation procedures.

- **Conduct incident response simulations**: Conduct regular incident response simulations to test the effectiveness of the incident response plan and identify areas for improvement.

- **Use automation to speed up incident response**: Use automation to speed up incident response and reduce the impact of security incidents – for example, automate the creation of backups, snapshots, and recovery procedures.

- **Implement real-time monitoring and alerts**: Implement real-time monitoring and alerts for AWS infrastructure and services to detect security incidents as soon as possible.

- **Follow AWS security best practices**: Follow AWS security best practices, such as implementing IAM policies and monitoring logs, to help prevent security incidents from occurring in the first place.

By following these best practices for incident response in AWS, organizations can improve their ability to detect, respond to, and recover from security incidents in a timely and effective manner. Additionally, organizations should regularly review and update their incident response plan to ensure it remains effective in the face of new and emerging security threats.

In this section, we discussed best practices for governing security in AWS, including leveraging AWS security services and features, maintaining compliance with various security standards and certifications, implementing governance frameworks such as the AWS Well-Architected Framework and AWS Security Hub, monitoring and logging for AWS security, and incident response for AWS security. By following these best practices, organizations can ensure that their AWS infrastructure is designed, implemented, and operated securely and in compliance with relevant standards and regulations.

In the next section, we'll discuss how to achieve these security best practices using Terraform. We'll explore how to use Terraform to govern security in AWS infrastructure, including implementing IAM policies, creating secure network architectures, and automating compliance checks. By the end of the next section, you'll have a solid understanding of how to use Terraform to implement and maintain secure infrastructure in AWS.

How to build secure infrastructure in Terraform

Terraform is an IaC tool that enables organizations to define and manage IaC. By using Terraform to build and manage infrastructure in AWS, organizations can achieve greater agility, scalability, and security. In this section, we'll explore best practices for building secure infrastructure in Terraform.

By following these best practices, organizations can build secure and compliant infrastructure in AWS using Terraform.

Implementing least privilege using IAM policies

IAM is a service provided by AWS that enables organizations to manage access to AWS resources and services. IAM policies are a key component of IAM that specify the permissions that are granted to AWS users, groups, and roles. Implementing least privilege using IAM policies means granting users, groups, and roles the minimum permissions required to perform their tasks. This can help reduce the risk of unauthorized access to AWS resources and services. Here are some best practices for implementing least privilege using IAM policies in Terraform:

- **Use IAM roles instead of IAM users**: IAM roles are a more secure way to grant access to AWS resources than IAM users. IAM roles can be assigned to AWS services or AWS EC2 instances, allowing for secure access without the need for long-term credentials.

- **Use the principle of least privilege**: Use the principle of least privilege to grant users, groups, and roles the minimum permissions required to perform their tasks. Avoid using policies that grant blanket permissions to AWS resources or services.

- **Use IAM policy conditions**: Use IAM policy conditions to specify additional conditions that must be met before access is granted to AWS resources or services. For example, you can require that access is only granted from a specific IP address or during a specific period.

- **Use Terraform modules for managing IAM policies**: Use Terraform modules to manage IAM policies and ensure that they are applied consistently across all AWS accounts and resources. This can help reduce the risk of misconfigurations and security vulnerabilities.

By utilizing these best practices when implementing least privilege using IAM policies in Terraform, organizations can ensure that their AWS resources and services are accessed only by authorized users and with the minimum permissions required to perform their tasks. Additionally, organizations should regularly review and update their IAM policies to ensure that they remain effective for emerging security threats.

Creating secure network architectures

Network security is a critical component of a secure infrastructure in AWS. By creating secure network architectures in Terraform, organizations can protect their infrastructure and data against network-based attacks. Here are some best practices for creating secure network architectures in Terraform:

- **Use VPCs to isolate resources**: Use Amazon **Virtual Private Cloud** (**VPC**) to isolate AWS resources and services from the public internet. VPCs enable organizations to create a private network within AWS and control access to resources using network security groups and ACLs.

- **Implement multiple layers of security**: Implement multiple layers of security to protect resources against network-based attacks. For example, use a public subnet for resources that need to be accessible from the internet, but place them behind an **Elastic Load Balancer** (**ELB**) and use security groups to control access. Use a private subnet for resources that should not be accessible from the internet, such as databases or other sensitive data.

- **Use AWS security services**: Use AWS security services, such as AWS WAF and AWS Shield, to protect against network-based attacks. WAF provides customizable web security rules to protect against common web exploits, while Shield provides continuous monitoring and automatic protection against **distributed denial-of-service** (**DDoS**) attacks.

- **Implement secure remote access**: Implement secure remote access to AWS resources using a bastion host or **virtual private network** (**VPN**). These solutions enable authorized users to access AWS resources securely from remote locations.

- **Use Terraform modules for network configuration**: Use Terraform modules to manage network configuration and ensure that it is applied consistently across all AWS resources and accounts. This can help reduce the risk of misconfigurations and security vulnerabilities.

By implementing these best practices for creating secure network architectures in Terraform, organizations can ensure that their AWS resources and services are protected against network-based attacks and that access is controlled and monitored.

Automating compliance checks

Compliance checks are an important aspect of maintaining a secure and compliant infrastructure in AWS. By automating compliance checks in Terraform, organizations can ensure that their infrastructure complies with relevant standards and regulations. Here are some best practices for automating compliance checks in Terraform:

- **Use AWS Config Rules**: AWS Config Rules is a service provided by AWS that enables organizations to define rules that evaluate the configuration of AWS resources and services against a set of predefined or custom rules. By using AWS Config Rules, organizations can automate compliance checks and detect non-compliant resources.

- **Implement continuous compliance monitoring**: Implement continuous compliance monitoring to detect non-compliant resources in real time. Continuous compliance monitoring can help organizations identify and remediate compliance issues before they become security vulnerabilities.

- **Use Terraform modules for compliance configuration**: Use Terraform modules to manage compliance configuration and ensure that it is applied consistently across all AWS resources and accounts. This can help reduce the risk of misconfigurations and security vulnerabilities.

- **Integrate compliance checks into the CI/CD pipeline**: Integrate compliance checks into the CI/CD pipeline to ensure that compliance checks are performed automatically as part of the infrastructure deployment process. This can help prevent non-compliant resources from being deployed in the first place.

By automating compliance checks in Terraform, organizations can ensure that their infrastructure is compliant with relevant standards and regulations and that non-compliant resources are detected and remediated promptly. Additionally, organizations should regularly review and update their compliance checks to ensure they remain effective in the face of new security threats.

Storing secrets securely

Storing secrets securely is a critical component of maintaining a secure infrastructure in AWS. Secrets such as API keys, passwords, and other sensitive information should be protected from unauthorized access. Here are some best practices for storing secrets securely in Terraform:

- **Use AWS Secrets Manager**: AWS Secrets Manager is a service provided by AWS that enables organizations to securely store and manage secrets, such as database credentials, API keys, and passwords. AWS Secrets Manager provides automatic rotation of secrets, audit logging, and fine-grained access controls.

- **Use AWS KMS**: AWS KMS is a service provided by AWS that enables organizations to create and control the encryption keys used to encrypt data. Use AWS KMS to encrypt secrets stored in AWS Secrets Manager and other storage solutions.

- **Avoid hardcoding secrets in Terraform code**: Avoid hardcoding secrets in Terraform code or storing them in plain text files. Instead, use environment variables or external storage solutions such as AWS Secrets Manager to store and manage secrets securely.

- **Use Terraform workspaces**: Use Terraform workspaces to manage secrets for different environments, such as development, staging, and production. This can help ensure that secrets are kept separate and secure for each environment.

By implementing these best practices for storing secrets securely in Terraform, organizations can protect sensitive information from unauthorized access and ensure that secrets are managed securely across all environments. To stay ahead of new and emerging security threats, it's essential for organizations to regularly assess and update their secret management practices.

Managing Terraform state

Terraform state is a critical component of managing IaC in AWS. Terraform state represents the current state of the infrastructure defined in Terraform code and is used to plan, apply, and modify infrastructure changes. Managing Terraform state securely is important to maintain the integrity and security of the infrastructure. Here are some best practices for managing Terraform state in Terraform:

- **Store Terraform state remotely**: Store Terraform state remotely in a secure and durable location, such as an Amazon S3 bucket or an external service such as HashiCorp's Terraform Cloud. Storing Terraform state remotely ensures that it is accessible by multiple team members and is not lost if the local machine is lost or damaged.

- **Use state locking**: Use state locking to prevent concurrent modifications to the Terraform state. State locking ensures that only one user or process can modify the state at a time, preventing conflicts and data corruption.

- **Encrypt Terraform state**: Encrypt Terraform state using a strong encryption algorithm and key management solution, such as AWS KMS. Encrypting Terraform state protects it from unauthorized access and ensures that sensitive data is kept confidential.

- **Regularly back up Terraform state**: Regularly back up Terraform state to prevent data loss in the event of a disaster. Backups should be stored in a secure and durable location, such as an Amazon S3 bucket versioning enabled.

Managing Terraform state securely is critical to maintaining the integrity and security of the infrastructure defined in Terraform code. By following these best practices for managing Terraform state in Terraform, organizations can ensure that their infrastructure is managed securely and that data is protected from unauthorized access and data loss.

Security and Terraform

Terraform is a powerful tool for managing IaC in AWS, but it also introduces new security challenges. In this section, we'll explore how Terraform can be used to enhance the security of AWS infrastructure, as well as some potential security risks and how to mitigate them.

By understanding the security implications of using Terraform in AWS and implementing best practices for secure Terraform usage, organizations can leverage the full potential of Terraform while maintaining a secure infrastructure.

The security benefits of using Terraform

Terraform offers several security benefits when managing IaC in AWS. Here are some of the key security benefits of using Terraform:

- **Consistent configuration**: Terraform enables organizations to define IaC, ensuring that the infrastructure is deployed in a consistent and repeatable manner. This can help reduce the risk of misconfigurations and security vulnerabilities.

- **Infrastructure versioning**: Terraform enables organizations to version their infrastructure, making it easier to track changes and revert to previous versions if necessary. This can help reduce the risk of unauthorized changes and maintain the integrity of the infrastructure.

- **Automation**: Terraform enables organizations to automate infrastructure deployment and management, reducing the risk of human error and increasing the speed of deployment. This can help reduce the risk of misconfigurations and security vulnerabilities.

- **Collaboration**: Terraform enables collaboration between teams, making it easier to manage IaC securely and efficiently. Collaboration can help reduce the risk of misconfigurations and security vulnerabilities by ensuring that infrastructure changes are reviewed and approved by multiple team members.

By leveraging the security benefits of using Terraform, organizations can deploy and manage infrastructure securely and efficiently. Additionally, by implementing best practices for using Terraform securely and mitigating common security risks, organizations can ensure that their infrastructure is protected against unauthorized access and other security threats.

Best practices for using Terraform securely

While Terraform offers several security benefits, it also introduces new security risks if not used securely. Here are some best practices for using Terraform securely in AWS:

- **Use least privilege**: Use least privilege when configuring IAM roles and policies for Terraform. Only grant permissions to the AWS resources and services that Terraform requires to manage infrastructure.

- **Store secrets securely**: Store secrets securely, such as API keys and passwords, using AWS Secrets Manager or other secure storage solutions. Avoid hardcoding secrets in Terraform code or storing them in plain text files.

- **Manage Terraform state securely**: Manage Terraform state securely by storing it remotely in a secure and durable location, using state locking to prevent concurrent modifications, encrypting it using a strong encryption algorithm and key management solution, and regularly backing it up.

- **Implement version control**: Implement version control for Terraform code using a version control system such as Git. Version control enables organizations to track changes to Terraform code, identify who made the changes, and revert to previous versions if necessary.

- **Audit and monitor Terraform usage**: Audit and monitor Terraform usage to detect unauthorized access and potential security threats. Use AWS CloudTrail to log all Terraform API calls and monitor CloudTrail logs for suspicious activity.

By implementing these best practices for using Terraform securely in AWS, organizations can reduce the risk of security vulnerabilities and protect their infrastructure against unauthorized access and other security threats. Additionally, organizations should regularly review and update their Terraform security practices to ensure that they remain effective in the face of new and emerging security threats.

Common security risks with Terraform and how to mitigate them

While Terraform offers several security benefits, it also introduces new security risks if not used securely. Here are some common security risks with Terraform and how to mitigate them:

- **Misconfigured IAM roles and policies**: Misconfigured IAM roles and policies can lead to unauthorized access and data breaches. To mitigate this risk, use least privilege when configuring IAM roles and policies for Terraform and regularly review and update them to ensure that they remain effective.

- **Storing secrets insecurely**: Storing secrets insecurely, such as API keys and passwords, can lead to unauthorized access and data breaches. To mitigate this risk, store secrets securely using AWS Secrets Manager or other secure storage solutions and avoid hardcoding secrets in Terraform code or storing them in plain text files.

- **Insecure Terraform code**: Insecure Terraform code can lead to misconfigurations and security vulnerabilities. To mitigate this risk, use best practices for writing secure Terraform code, such as avoiding hardcoding sensitive information, using modules for code reuse, and using parameterized values for sensitive data.

- **Misconfigured Terraform state management**: Misconfigured Terraform state management can lead to data corruption and unauthorized access. To mitigate this risk, use best practices for managing Terraform state, such as storing it remotely in a secure and durable location, using state locking to prevent concurrent modifications, encrypting it using a strong encryption algorithm and key management solution, and regularly backing it up.

In conclusion, Terraform offers many benefits for managing IaC in AWS, but it also introduces new security risks. By understanding the common security risks associated with Terraform and implementing best practices for using Terraform securely, organizations can reduce the risk of security vulnerabilities and protect their infrastructure against unauthorized access and other security threats. Regularly reviewing and updating Terraform security practices is important to ensure that they remain effective in the face of new and emerging security threats.

Security and IaC operations

IaC operations are critical to ensuring the security and stability of AWS infrastructure. In this section, we'll explore the security implications of IaC operations in AWS.

By understanding the security implications of IaC operations in AWS and implementing best practices for secure IaC operations, organizations can ensure the ongoing security and stability of their infrastructure.

IaC pipeline security

IaC pipelines are used to automate the build, test, and deployment of IaC in AWS. It is important to ensure the security of IaC pipelines to prevent unauthorized access and modifications to the code, as well as to protect against potential security vulnerabilities. Here are some best practices for securing IaC pipelines in AWS:

- **Use version control**: Use version control for IaC code to enable tracking of changes, collaboration, and accountability. Consider using a version control system such as Git to store code.

- **Implement access controls**: Implement access controls to restrict access to the IaC pipeline and associated AWS resources. Use AWS IAM roles and policies to limit access to only authorized users and services.

- **Securely store artifacts**: Securely store artifacts generated during the IaC pipeline, such as compiled code, test reports, and configuration files. Consider using an artifact repository such as Amazon S3 to store artifacts.

- **Enable encryption**: Enable encryption for IaC pipeline resources, such as build servers and artifact repositories. Consider using AWS KMS to manage encryption keys.

- **Implement continuous monitoring**: Implement continuous monitoring of the IaC pipeline to detect potential security vulnerabilities or unauthorized access. Consider using AWS CloudTrail to monitor API calls and AWS Config to monitor resource configurations.

By implementing these best practices for securing IaC pipelines in AWS, organizations can ensure that their IaC is deployed securely and remains protected against unauthorized access and other security threats.

Securing build and deployment processes

Securing the build and deployment processes for IaC in AWS is critical to maintaining the security and stability of the infrastructure. Here are some best practices for securing the build and deployment processes:

- **Implement secure coding practices**: Implement secure coding practices to prevent the introduction of security vulnerabilities into the infrastructure code. Examples of secure coding practices include validating user input, using parameterized values for sensitive data, and avoiding hardcoded secrets in code.

- **Use code analysis tools**: Use code analysis tools to identify potential security vulnerabilities in infrastructure code. Consider using tools such as AWS CodeGuru or third-party code analysis tools.

- **Implement testing and validation**: Implement testing and validation of infrastructure changes before deployment to detect potential security vulnerabilities or misconfigurations. Consider using tools such as AWS CodeBuild or GitHub Actions for automated testing and validation.

- **Enable auditing and logging**: Enable auditing and logging of build and deployment processes to detect potential security threats or unauthorized access. Consider using AWS CloudTrail to monitor API calls and AWS Config to monitor resource configurations.

- **Use deployment pipelines**: Use deployment pipelines to automate the deployment of infrastructure changes to reduce the risk of human error and ensure consistency. Consider using AWS CodePipeline or other deployment pipeline tools.

By implementing these best practices for securing the build and deployment processes for IaC in AWS, organizations can ensure that their infrastructure remains secure and stable.

Securely managing secrets in IaC pipelines

Secrets management is critical to the security of IaC pipelines in AWS. Secrets, such as API keys, passwords, and certificates, must be managed securely to prevent unauthorized access and data breaches. Here are some best practices for securely managing secrets in IaC pipelines:

- **Use secrets management tools**: Use secrets management tools, such as AWS Secrets Manager, to store and manage secrets securely. Secrets should be encrypted at rest and in transit, and access should be restricted to authorized users and services.

- **Use IAM roles and policies**: Use AWS IAM roles and policies to control access to secrets. Access should be limited to only the resources and services that require it.

- **Avoid hardcoding secrets**: Avoid hardcoding secrets in IaC code or storing them in plain text files. Instead, use environment variables or secrets management tools to retrieve the secrets.

- **Implement audit logging**: Implement audit logging of secret access and usage to detect potential security threats or unauthorized access. Consider using AWS CloudTrail to log secret access.

- **Rotate secrets regularly**: Rotate secrets regularly to reduce the risk of data breaches or unauthorized access. Consider using AWS Secrets Manager to automate secret rotation.

Testing and validating infrastructure changes

Testing and validating infrastructure changes is critical to ensuring the security and stability of infrastructure in AWS. Here are some best practices for testing and validating infrastructure changes:

- **Use infrastructure testing tools**: Use infrastructure testing tools, such as Terraform's `plan` command, to test changes before deployment. This can help identify potential security vulnerabilities or misconfigurations.

- **Implement code reviews**: Implement code reviews to identify potential security vulnerabilities or misconfigurations in infrastructure code. Code reviews can also help ensure that code follows best practices and is consistent with organizational standards.

- **Conduct regular vulnerability assessments**: Conduct regular vulnerability assessments of infrastructure code to identify potential security vulnerabilities. Consider using third-party vulnerability assessment tools to supplement internal assessments.

- **Conduct regular security audits**: Conduct regular security audits of infrastructure code to identify potential security threats or unauthorized access. Consider using AWS Config Rules or third-party security audit tools.

Best practices for secure IaC operations

Implementing best practices for secure IaC operations is critical to ensuring the security and stability of infrastructure in AWS. Here are some best practices for secure IaC operations:

- **Follow the principle of least privilege**: Follow the principle of least privilege when granting access to IaC resources and services. Use AWS IAM roles and policies to restrict access to only the resources and services that are required.

- **Implement change management**: Implement change management processes to ensure infrastructure changes are reviewed, tested, and approved before deployment. Consider using AWS Service Catalog or other change management tools.

- **Use IaC templates**: Use IaC templates to ensure consistency and repeatability in infrastructure deployment. Consider using Terraform templates or modules.

- **Implement security automation**: Implement security automation to identify potential security vulnerabilities or misconfigurations in infrastructure code. Consider using AWS Config Rules or third-party security automation tools.

- **Train the team on security best practices**: Train the team on security best practices to ensure that they are aware of potential security threats and know how to mitigate them. Regularly conduct security awareness training for all staff members.

Summary

In this chapter, we explored the importance of security in infrastructure and how to build secure infrastructure in AWS using Terraform. We discussed the basic principles of infrastructure security, types of security measures for infrastructure, and the role of governance in infrastructure security.

We also covered best practices for governing security in AWS, including AWS security services and features, security compliance and certifications, security governance frameworks, monitoring and logging for security, and incident response for security.

Additionally, we explored best practices for building secure infrastructure in Terraform, including implementing least privilege using IAM policies, creating secure network architectures, automating compliance checks, securely managing secrets, and managing Terraform state.

Then, we delved into the security benefits of using Terraform, best practices for using Terraform securely, and common security risks with Terraform and how to mitigate them.

Finally, we discussed the security implications of IaC operations in AWS, including IaC pipeline security, securing build and deployment processes, securely managing secrets in IaC pipelines, testing and validating infrastructure changes, and best practices for secure IaC operations.

Overall, this chapter emphasized the critical importance of security in infrastructure and the various best practices that organizations should follow to ensure the ongoing security and stability of their infrastructure in AWS. By implementing these best practices, organizations can minimize the risk of potential security threats, data breaches, and other security incidents and protect their infrastructure and data in AWS.

In the next chapter, we will learn how to design and develop infrastructure for perfection and how we can maintain it over time.

Perfecting AWS Infrastructure with Terraform

"What does it mean to have perfect infrastructure?" In this final chapter, we will explore what it means to achieve perfection in cloud infrastructure and how to design, develop, and continuously improve it. We will also delve into building **service-level agreements (SLAs)**, **service-level indicators (SLIs)**, and **service-level objectives (SLOs)** with **site reliability engineering (SRE)** principles. Additionally, we will cover how to run operations with Terraform, including monitoring, observability, logging, debugging, and building repeatable environments. By the end of this chapter, you will have gained a comprehensive understanding of what it takes to achieve perfection in your AWS infrastructure and how to maintain it over time.

We will cover the following main topics:

- What does perfect mean in cloud infrastructure?
- How to design and develop infrastructure for perfection
- Continuously improving cloud infrastructure
- Building SLAs/SLIs/SLOs with SRE principles
- How to run operations with Terraform

What does perfect mean in cloud infrastructure?

When it comes to cloud infrastructure, achieving perfection means designing and building an environment that meets the needs of all stakeholders, is highly available, secure, scalable, and efficient, and is continuously improving over time. In this section, we will explore what perfection means in cloud infrastructure and provide some guidelines for achieving it.

Meeting stakeholder needs

Meeting stakeholder needs is a critical aspect of designing and building perfect cloud infrastructure with Terraform. It involves understanding the requirements and expectations of all stakeholders, including customers, users, managers, and technical teams, and developing solutions that meet their needs.

To meet stakeholder needs, it is essential to engage in effective communication and collaboration. This includes regular meetings, feedback sessions, and open communication channels to discuss requirements, provide updates, and gather feedback.

In addition to communication, it is important to have a clear understanding of stakeholder priorities and goals. This involves identifying critical success factors, such as performance, scalability, security, and cost-effectiveness, and developing solutions that prioritize these factors.

Furthermore, it is essential to have a deep understanding of the business and technical requirements of each stakeholder group. For example, customer requirements may focus on user experience and reliability, while technical teams may prioritize automation and scalability.

When designing and building perfect cloud infrastructure with Terraform, it is important to keep stakeholder needs in mind throughout the process. This involves continuously iterating and improving solutions based on stakeholder feedback and evolving business requirements.

By meeting stakeholder needs, you can ensure that your cloud infrastructure meets the expectations of all stakeholders, delivers maximum value, and supports the success of your business.

High availability

High availability is a critical factor in ensuring that your infrastructure can meet the demands of your users and customers. It refers to the ability of a system or application to remain operational and accessible in the event of hardware or software failures, network disruptions, or other unforeseen events. Achieving high availability requires careful planning and design, as well as the use of appropriate technologies and strategies.

One key aspect of achieving high availability is redundancy. This involves deploying multiple instances of your application or service across different availability zones or regions. By spreading your workload across multiple instances, you can ensure that if one instance fails or becomes unavailable, traffic can be routed to another instance, minimizing downtime and maintaining service availability.

Another important strategy for achieving high availability is load balancing. Load balancers distribute traffic across multiple instances of your application or service, helping to ensure that no single instance becomes overloaded and that traffic can be automatically routed to healthy instances in the event of a failure.

In addition to redundancy and load balancing, other strategies for achieving high availability are as follows:

- **Implementing automatic failover**: This involves automatically shifting traffic to healthy instances in the event of a failure, without requiring manual intervention

- **Monitoring and alerting**: Implementing monitoring and alerting systems can help you quickly detect and respond to issues before they become major problems

- **Disaster recovery planning**: Creating a disaster recovery plan can help you quickly recover from major failures or disasters and minimize downtime

By implementing these strategies and others, you can help ensure that your infrastructure remains highly available and resilient, even in the face of unexpected challenges.

Security

Security is a critical aspect of any cloud infrastructure, and it is essential to design and implement security measures that protect against potential threats. When designing and deploying infrastructure with Terraform in AWS, it is crucial to follow AWS security best practices and ensure that all resources are properly secured.

One of the first steps to achieving a secure infrastructure is to establish **Identity and Access Management (IAM)** policies that control who can access the resources and what actions they can perform. IAM policies should be designed with the principle of least privilege, which means that users should only have access to the resources that are necessary for them to perform their duties.

Another key aspect of security is network security. In AWS, network security can be achieved through the use of security groups and network **access control lists (ACLs)** to control traffic flow between resources. Security groups are stateful firewalls that control inbound and outbound traffic for an instance, while network ACLs are stateless and can control traffic at the subnet level.

Encryption is also essential for securing data in transit and at rest. AWS provides various encryption options, including server-side encryption with Amazon S3, client-side encryption for Amazon S3 and Amazon EBS, and end-to-end encryption with AWS **Key Management Service (KMS)**.

In addition to these measures, it is crucial to implement monitoring and logging to detect and respond to potential security threats. AWS provides various monitoring and logging tools, such as Amazon CloudWatch, AWS Config, and AWS CloudTrail, that can be used to monitor and track activity across your infrastructure.

Overall, implementing security measures in your Terraform infrastructure requires a comprehensive approach that covers all aspects of the cloud environment, including IAM, network security, encryption, and monitoring. By following AWS security best practices and using the appropriate security tools and services, you can ensure that your infrastructure is secure and protected against potential threats.

Scalability

Scalability is a critical aspect of cloud infrastructure design as it allows you to grow your resources as your needs increase without disrupting your existing systems. Scalability ensures that your applications and services can handle increasing traffic and workloads while maintaining performance and availability.

Designing for scalability requires carefully considering various factors, including workload patterns, data storage needs, and network traffic. The goal is to create a flexible and resilient infrastructure that can easily accommodate growth without impacting performance or user experience.

Here are some key considerations for designing scalable infrastructure:

- **Elasticity**: The ability to dynamically scale resources up or down based on demand
- **Load balancing**: This involves distributing traffic across multiple instances or resources to avoid overloading any one resource
- **Autoscaling**: This involves automatically adjusting resource capacity in response to changes in demand
- **Database scalability**: This involves choosing the right database architecture and scaling strategy to ensure that your data storage can grow along with your infrastructure
- **Network scalability**: This involves ensuring that your network can handle increasing traffic and load, and scaling resources accordingly

Scalability is essential for modern cloud infrastructure as it allows businesses to keep pace with changing demands and stay competitive. With careful planning and the right tools, you can design and deploy a highly scalable infrastructure that can grow and evolve with your needs.

Efficiency

Efficiency is a crucial aspect of cloud infrastructure, and it can have a significant impact on cost, performance, and reliability. When designing and implementing infrastructure with Terraform, it's essential to consider efficiency from the outset. In this section, we'll explore the key factors to consider when building efficient infrastructure with Terraform.

Efficient use of resources

Efficient use of resources is critical to achieving cost-effective and high-performing infrastructure. When building infrastructure with Terraform, it's essential to consider the appropriate sizing of resources, such as EC2 instances, RDS databases, and storage volumes. This involves selecting the right instance type, storage type, and amount of resources for the workload.

One way to achieve efficient use of resources is to implement autoscaling policies. Autoscaling allows you to scale resources up or down based on changes in demand, ensuring that you're using only the resources that you need at any given time.

Optimizing network performance

Network performance is another critical factor in infrastructure efficiency. When building infrastructure with Terraform, it's important to optimize network performance by selecting the appropriate network architecture, such as VPCs, subnets, and security groups. This involves considering factors such as latency, bandwidth, and security requirements.

One way to optimize network performance is to implement **content delivery networks** (**CDNs**) and edge caching. CDNs and edge caching allow you to distribute content closer to the end users, reducing latency and improving performance.

Automation and continuous improvement

Efficiency also involves automation and continuous improvement. When building infrastructure with Terraform, it's important to automate repetitive tasks, such as deployment, testing, and monitoring. This allows you to focus on more critical tasks, such as development and innovation.

Continuous improvement involves monitoring and analyzing infrastructure performance, identifying areas for improvement, and implementing changes to optimize performance and efficiency.

Efficiency is a crucial aspect of cloud infrastructure, and it's essential to consider it from the outset when building infrastructure with Terraform. By optimizing resource usage, network performance, and automation, you can achieve cost-effective, high-performing infrastructure that meets your business needs.

Continuous improvement

Continuous improvement is an essential part of creating and maintaining perfect cloud infrastructure. It involves constantly evaluating and refining your infrastructure to ensure that it is operating at peak efficiency and meeting the needs of your stakeholders. To achieve continuous improvement, you need to establish a culture of continuous learning and experimentation and embrace tools and techniques that can help you measure and analyze the performance of your infrastructure.

One important tool for continuous improvement is monitoring. By monitoring your infrastructure, you can track performance metrics, identify potential issues, and proactively address them before they become critical. You can use tools such as AWS CloudWatch to monitor your AWS resources and applications in real time and set up alerts to notify you when specific events occur.

Another important technique for continuous improvement is automation. By automating common tasks and processes, you can reduce the likelihood of human error and improve efficiency. Terraform provides a powerful platform for automating infrastructure tasks, allowing you to define and manage infrastructure as code.

In addition to monitoring and automation, you can also leverage feedback from stakeholders to identify areas for improvement. Regularly soliciting feedback from stakeholders can help you identify pain points, bottlenecks, and other areas that can be improved. This feedback can be used to inform your continuous improvement efforts and guide your infrastructure development.

Ultimately, achieving perfect cloud infrastructure requires a commitment to continuous improvement. By embracing tools and techniques that promote monitoring, automation, and stakeholder feedback, you can ensure that your infrastructure is always operating at peak efficiency and meeting the needs of your organization.

How to design and develop infrastructure for perfection

To achieve perfection in AWS infrastructure, it is crucial to approach infrastructure design and development with a comprehensive focus on meeting stakeholder requirements, ensuring high availability and security, enabling scalability, and optimizing efficiency. In this section, we will explore the critical factors that go into designing and developing infrastructure that meets these demands, while leveraging the power of **Infrastructure as Code (IaC)** with Terraform.

Defining infrastructure requirements

One of the first steps in developing perfect infrastructure is defining the requirements of all stakeholders, including developers, operations, and management teams. This can involve developing a comprehensive understanding of the technical and business needs of each group and incorporating them into the overall design and development strategy. Using IaC tools such as Terraform can help facilitate this process by allowing stakeholders to collaborate on a shared code base and visualize the infrastructure design in a way that is easily understandable to all parties.

For example, a stakeholder may require the infrastructure to have high availability, low latency, and quick recovery time in case of a disaster. Another stakeholder may require the infrastructure to be cost-effective and scalable to handle peak traffic. By clearly defining these requirements, the design team can create a roadmap for developing the infrastructure and ensure that all parties are working toward a common goal.

Furthermore, by utilizing IaC tools such as Terraform, infrastructure requirements can be codified, version-controlled, and tested just like any other software code. This approach allows for more efficient and accurate communication between stakeholders, as changes to the infrastructure can be made via code changes and tracked through version control. It also provides the ability to automate the deployment of infrastructure changes, reducing the risk of human error and improving the speed of deployment.

Establishing a design framework

Once the infrastructure requirements have been defined, the next step is to establish a design framework that will guide the development process. This involves defining the architectural principles, standards, and patterns that will be used to build the infrastructure. Using IaC tools such as Terraform can help establish a consistent design framework and ensure that the infrastructure is built to a high standard.

Some important considerations when establishing a design framework are as follows:

- Choosing the appropriate architecture style for the application or workload, such as microservices, serverless, or monolithic
- Selecting the appropriate AWS services and components to build the infrastructure, based on the defined requirements and design principles
- Defining the relationships and dependencies between different components of the infrastructure, to ensure that they work together smoothly and efficiently
- Developing a set of design patterns and best practices to be used throughout the development process, to ensure consistency and maintainability
- Using IaC tools such as Terraform to define the infrastructure in code, which provides version control, reproducibility, and consistency

By establishing a clear design framework, developers can ensure that the infrastructure is built to a high standard and that it meets the requirements of all stakeholders.

Implementing best practices

Once the design framework has been established, it's important to follow industry best practices for infrastructure development and deployment. This includes implementing security measures such as encryption, access controls, and IAM. IaC tools such as Terraform can help ensure consistent implementation of these practices across environments.

In addition, it's important to establish guidelines for code quality, testing, and review to ensure that the infrastructure is reliable and efficient. This can involve creating automated testing and deployment pipelines, setting up monitoring and alerting systems, and establishing disaster recovery and business continuity plans.

By implementing best practices for infrastructure development and deployment, teams can reduce the risk of security breaches, downtime, and other issues that can negatively impact the business. Terraform can be a valuable tool for implementing these best practices, as it enables teams to easily define and manage infrastructure in a consistent and repeatable manner.

Testing and validating the infrastructure

Once the infrastructure has been designed and implemented, it's essential to test and validate it thoroughly. This involves ensuring that the infrastructure meets the defined requirements, is secure, and is reliable.

Automated testing is crucial in ensuring that the infrastructure is functioning correctly. Tools such as Terraform can help automate the testing process by allowing you to define test cases and run them automatically. You can also use tools such as AWS CloudFormation to create templates for testing and validating infrastructure.

In addition to automated testing, it's also important to perform manual testing and validation. This can involve reviewing logs, monitoring system performance, and conducting security assessments. It's important to establish clear processes for testing and validation and ensure that all team members understand their roles and responsibilities in the process.

Here are some examples of scenarios for testing and validating infrastructure:

- **Testing disaster recovery procedures**: Simulate various failure scenarios, such as a server going down, and ensure that the infrastructure can recover without data loss or downtime

- **Load testing**: Simulate high-traffic scenarios and ensure that the infrastructure can handle the increased load without downtime or performance degradation

- **Security testing**: Perform vulnerability assessments and penetration testing to identify and address potential security risks

By thoroughly testing and validating the infrastructure, you can ensure that it meets the defined requirements, is secure, and is reliable.

Continuous integration and continuous deployment (CI/CD)

CI/CD is an essential aspect of modern infrastructure development. With CI/CD, changes to the infrastructure are automatically built, tested, and deployed to production environments. This helps ensure that the infrastructure is always up to date and free of errors.

To implement CI/CD, it is necessary to establish an automated pipeline that integrates with version control systems such as Git, automated testing tools, and infrastructure deployment tools such as Terraform. This pipeline should be designed to ensure that each change to the infrastructure is thoroughly tested before being deployed to production.

One common approach to implementing CI/CD with Terraform is to use a CI tool such as Jenkins, GitHub Actions, or Terraform Cloud to automate the build and deployment process. The pipeline would include steps to clone the Terraform code repository, validate and test the code, and deploy the changes to the target environment.

Automated testing is an essential aspect of a robust CI/CD pipeline. Terraform provides several options for testing infrastructure code, including unit testing, integration testing, and acceptance testing. Unit testing involves testing individual modules or resources, while integration testing tests the interaction between modules or resources. Acceptance testing involves testing the entire infrastructure against defined requirements.

To ensure that the infrastructure changes are deployed to production only when they have been thoroughly tested and meet defined quality standards, it is also essential to establish a process for review and approval of code changes. Code review should involve a peer review process where other team members review the changes and provide feedback. Approvals should be granted only after thorough testing and validation.

Continuously improving cloud infrastructure

Continuously improving cloud infrastructure is a crucial component of ensuring that it remains optimal and efficient over time. It involves implementing processes and strategies that help identify areas for improvement, addressing them, and tracking the effectiveness of changes made. In this section, we will discuss the key concepts and strategies for continuously improving cloud infrastructure. We will also explore how Terraform can be used as a powerful tool to automate the implementation of changes and to help ensure that improvements are made in a consistent and repeatable manner.

Monitoring and logging

One of the key components of continuous improvement is monitoring and logging. This involves implementing a comprehensive monitoring and logging system to track the performance and health of your infrastructure and applications. This can include metrics such as CPU and memory usage, network traffic, and application-specific metrics.

Terraform can play an important role in setting up monitoring and logging systems by deploying infrastructure resources such as Amazon CloudWatch, which can be used to monitor infrastructure and application logs. Amazon CloudWatch also provides a range of dashboards and alerts to help you track the health and performance of your infrastructure in real time.

Other tools and services that can be used for monitoring and logging include Elasticsearch, Kibana, Grafana, and Fluentd. These tools can be used to collect, analyze, and visualize log data, as well as provide alerts for potential issues.

By monitoring and logging your infrastructure and applications, you can proactively identify and resolve issues, optimize performance, and continuously improve your overall infrastructure.

Alerting and notification

In addition to monitoring and logging, alerting and notification are critical components of continuously improving cloud infrastructure. This involves setting up alerts for specific metrics or events that could indicate an issue with the infrastructure, such as high CPU usage or low disk space. These alerts can be configured to trigger notifications to relevant stakeholders, such as operations or development teams, to ensure that the issue is addressed as quickly as possible.

Terraform can help with alerting and notification by allowing for the automated configuration of monitoring tools such as CloudWatch or Datadog, as well as setting up the necessary alerts and notifications. Terraform also enables the use of tools such as PagerDuty or Slack to ensure that notifications are sent to the appropriate channels and stakeholders. By leveraging Terraform to automate alerting and notification, organizations can ensure that their infrastructure is continuously monitored and any issues are quickly addressed.

Capacity planning and management

This aspect of continuous improvement involves analyzing current usage patterns and predicting future usage trends to ensure that the infrastructure can handle the expected load. It is important to monitor resource utilization and plan for additional capacity as needed to maintain high availability and prevent performance issues. With Terraform, capacity planning and management can be automated through the use of auto-scaling groups and the ability to easily adjust resource allocation in response to changing demand. This can help ensure that the infrastructure is always able to handle the workload and minimize downtime or performance issues. Additionally, capacity planning and management can help optimize costs by ensuring that resources are only allocated as needed, reducing waste and unnecessary spending.

Cost optimization and management

One important aspect of continuously improving cloud infrastructure is ensuring cost efficiency. This involves not only monitoring and managing costs but also implementing measures to optimize them. Terraform can play a crucial role in this process by allowing infrastructure to be designed and deployed cost-effectively.

One way to optimize costs is by implementing auto-scaling policies, which can automatically adjust resources based on demand. This prevents overprovisioning and reduces wasted resources, leading to cost savings.

Another way to optimize costs is by implementing reserved instances for services that have predictable usage patterns. Reserved instances offer discounted pricing in exchange for a commitment to use a specific number of resources over a certain period.

Additionally, leveraging AWS Cost Explorer and third-party tools can provide valuable insights into cost optimization opportunities. With Terraform, these optimizations can be incorporated into the infrastructure code to ensure continuous cost efficiency.

IaC review

IaC review is an important aspect of continuously improving cloud infrastructure. It involves regularly reviewing and updating the Terraform code to ensure that it is optimized, efficient, and follows best practices. The IaC review process can help identify and address issues such as unused resources, security vulnerabilities, and misconfigurations.

During the IaC review process, it is important to consider the following aspects:

- **Consistency**: Ensure that the Terraform code is consistent across all environments and follows a standard set of practices and conventions
- **Security**: Verify that the infrastructure is secure and meets all relevant compliance requirements
- **Scalability**: Ensure that the infrastructure can scale up or down as needed and that the Terraform code is optimized for performance
- **Cost-effectiveness**: Identify opportunities to optimize costs, such as by using reserved instances, spot instances, or auto-scaling

The IaC review process should be performed regularly, ideally as part of a CI/CD pipeline. This ensures that any changes to the infrastructure are reviewed and approved before they are deployed to production, helping to prevent issues and downtime.

Regular security audits and updates

Security is an essential aspect of any cloud infrastructure, and it is crucial to ensure that the infrastructure remains secure and up to date. Regular security audits can help identify potential security vulnerabilities and weaknesses in the infrastructure, while also providing recommendations for improving security.

In addition to security audits, regular updates to the infrastructure can also help to improve security. This includes updating software and patches to address any known vulnerabilities, as well as regularly reviewing and updating security policies and procedures. IaC can also play a significant role in ensuring security as it enables the automation of security controls and can help to ensure that the infrastructure is configured consistently and securely.

To stay on top of security updates and patches, it's important to have a defined process for managing security. This may include regular security scans and assessments, as well as establishing clear roles and responsibilities for security-related tasks. It's also essential to have a plan in place for responding to security incidents and breaches, including incident response procedures and communication plans.

Performance optimization and management

Performance optimization and management is another critical aspect of continuously improving cloud infrastructure. This involves monitoring and analyzing the performance of the infrastructure and applications to identify potential bottlenecks, areas for improvement, and opportunities for optimization.

To effectively manage and optimize performance, it is important to establish a performance baseline, set performance targets, and continually measure and analyze performance against these targets. This can involve collecting and analyzing data on metrics such as response times, latency, throughput, and error rates, and using this information to identify areas for improvement.

In terms of using Terraform for performance optimization and management, it can be used to provision and manage resources such as load balancers, auto-scaling groups, and performance monitoring tools. Terraform also allows for the automation of performance testing and optimization processes, enabling faster and more efficient testing and deployment of performance improvements.

Continuous improvement and iteration

Continuous improvement and iteration are crucial components of achieving perfect infrastructure in AWS. It involves regularly evaluating and identifying areas for improvement in the infrastructure and implementing changes to address these issues. This process helps ensure that the infrastructure remains efficient, secure, and scalable over time, and meets the evolving needs of the organization and its stakeholders. By adopting a continuous improvement and iteration approach, organizations can ensure that their infrastructure is always optimized to its fullest potential and that their investments in AWS are delivering maximum value.

Building SLAs/SLIs/SLOs with SRE principles

To ensure that cloud infrastructure meets the needs of its users and stakeholders, it is important to establish clear SLAs, SLIs, and SLOs that align with business goals. Additionally, it is important to utilize SRE principles to manage the service and maintain its reliability. This section will provide an overview of the concepts behind SLAs, SLIs, SLOs, and SRE, and how they can be integrated into the design and development of cloud infrastructure. By following these principles, organizations can improve the reliability and availability of their cloud services, and ensure that they are meeting the needs of their users and stakeholders.

What are SLAs, SLIs, and SLOs?

SLAs, SLIs, and SLOs are critical concepts in modern IT service management. SLAs are agreements between service providers and their customers that define the level of service that will be provided, including availability, response times, and other metrics. SLIs are metrics that are used to measure the performance of a service, while SLOs are specific targets for those metrics.

For example, an SLA might specify that a particular service must be available 99.99% of the time, with a maximum response time of 500 milliseconds. SLIs for this service might include availability and response time metrics, while SLOs would set specific targets for those metrics, such as 99.99% availability and a maximum response time of 500 milliseconds.

SRE is a set of principles and practices that focus on improving the reliability and availability of services. SRE teams work to ensure that services meet their SLAs, SLIs, and SLOs, and they use data and automation to continuously improve service reliability.

In this section, we will explore the principles of SRE and how to apply them to build and manage cloud infrastructure with Terraform. We will cover topics such as defining SLAs, SLIs, and SLOs, monitoring service performance, and using data and automation to improve service reliability.

Key principles of SRE

The key principles of SRE are a set of practices that are designed to improve the reliability and maintainability of software systems. The principles of SRE involve a focus on automation, monitoring, testing, and continuous improvement.

SRE teams are responsible for ensuring that systems are reliable, available, and scalable. They work closely with software developers to ensure that systems are designed with these principles in mind. SRE teams use monitoring tools to detect problems and take proactive measures to prevent system failures. They also conduct regular reviews of the system to identify areas for improvement.

Some of the key principles of SRE are as follows:

- **Automation**: SRE teams automate as many processes as possible, including testing, deployment, and monitoring. This helps to reduce errors and improve the efficiency of the system.
- **Monitoring**: SRE teams use monitoring tools to detect problems in the system. This helps them to identify issues before they become critical.
- **Testing**: SRE teams conduct regular tests of the system to identify any issues that may affect reliability. This helps them to proactively identify and resolve problems before they become critical.
- **Continuous improvement**: SRE teams are always looking for ways to improve the reliability and performance of the system. They conduct regular reviews of the system to identify areas for improvement.

Developing SLAs, SLIs, and SLOs

To implement SRE principles effectively, it is essential to develop SLAs, SLIs, and SLOs. SLAs define the formal agreement between the service provider and the customer, outlining the expectations for service delivery. SLIs and SLOs are used to measure the quality of service delivery and ensure that it meets the agreed-upon levels of performance.

SLIs are metrics that are used to measure the performance of the service. They provide a quantitative measurement of the quality of the service and are used to track whether the service is meeting the agreed-upon levels of performance. SLOs are specific, measurable goals for the quality of service. They define the expected level of service and the time frame in which it should be delivered. SLOs are used to ensure that the service meets the agreed-upon levels of performance.

Developing effective SLAs, SLIs, and SLOs requires a deep understanding of the service and the needs of the customer. It is important to identify the **key performance indicators** (**KPIs**) that are most important to the customer and develop metrics that measure them accurately. These metrics should be regularly reviewed and updated to ensure that they remain relevant to the evolving needs of the customer.

Using Terraform, it is possible to integrate monitoring and alerting tools to track SLIs and SLOs. This can help to ensure that the service meets the agreed-upon levels of performance and enables rapid response to any issues that arise. Additionally, Terraform can be used to automate the provisioning of resources required to meet SLOs, ensuring that the service can scale quickly to meet demand.

Measuring and monitoring metrics for SLIs and SLOs

Measuring and monitoring metrics for SLIs and SLOs is a critical aspect of ensuring that your infrastructure meets the defined performance standards. This involves selecting and tracking key metrics that provide insight into the health and performance of your infrastructure. Examples of metrics that can be used for SLIs and SLOs include response time, error rate, availability, and throughput.

Tools such as AWS CloudWatch and Prometheus can be used to collect and analyze these metrics in real time. Once you have established a baseline for these metrics, you can set thresholds for each metric that define when the infrastructure is meeting or failing to meet the defined performance standards. When a threshold is crossed, alerts can be triggered to notify the appropriate teams, allowing them to take action to resolve the issue and prevent further degradation of the infrastructure.

Using Terraform, you can also define and implement monitoring and alerting resources alongside your infrastructure code, ensuring that your monitoring and alerting are versioned, tested, and deployed along with your infrastructure. This allows for a more streamlined and integrated approach to monitoring and alerting, as well as making it easier to maintain and update these resources as your infrastructure evolves.

Using Terraform to enforce SLAs, SLIs, and SLOs

Using Terraform to enforce SLAs, SLIs, and SLOs involves creating and deploying infrastructure that meets specific performance requirements. This can involve defining and implementing IaC templates that incorporate specific metrics and monitoring tools, as well as configuring alerts and notifications to be triggered when performance metrics fall below certain thresholds.

By leveraging Terraform's ability to deploy and manage infrastructure at scale, teams can ensure that their infrastructure is consistently meeting performance requirements and providing a high level of reliability and availability. Terraform can also be used to automate the process of deploying updates and making infrastructure changes to continuously improve performance and optimize resource utilization.

To effectively use Terraform for SLA, SLI, and SLO enforcement, it is important to have a deep understanding of the underlying infrastructure and the specific requirements of the application or service being deployed. This requires close collaboration between development, operations, and management teams to ensure that the infrastructure is aligned with business goals and objectives.

Some of the key considerations when using Terraform for SLA, SLI, and SLO enforcement are as follows:

- Defining clear and measurable SLAs, SLIs, and SLOs

- Incorporating metrics and monitoring tools into IaC templates

- Configuring alerts and notifications to be triggered when performance metrics fall below certain thresholds

- Automating the process of deploying updates and making changes to the infrastructure

- Regularly reviewing and refining SLAs, SLIs, and SLOs to ensure they remain aligned with business goals and objectives

Best practices for managing SLAs, SLIs, and SLOs

Building and managing SLAs, SLIs, and SLOs is a critical component of ensuring the availability, reliability, and performance of your infrastructure. By defining and tracking these metrics, you can establish clear expectations for your users and stakeholders and hold yourself accountable for delivering the best possible experience. In this section, we will explore the key concepts and principles of SLOs and SLAs, and how you can use Terraform to enforce and manage these metrics in your AWS environment. We will also cover best practices for defining and measuring SLIs, and how to use this data to continuously improve your infrastructure.

Here are some best practices for managing SLAs, SLIs, and SLOs:

- **Collaborate with stakeholders**: Involve all stakeholders in the SLA, SLI, and SLO development process, including developers, operations, and management teams

- **Set realistic targets**: Ensure that the SLA, SLI, and SLO targets are achievable and based on business needs and user requirements

- **Define clear metrics**: Clearly define the metrics that will be used to measure SLI and SLO compliance

- **Monitor and measure**: Continuously monitor and measure the SLI and SLO metrics to ensure that they are being met

- **Automate where possible**: Use automation tools, such as Terraform, to help enforce SLAs, SLIs, and SLOs

- **Review and adjust**: Regularly review and adjust the SLA, SLI, and SLO targets based on changing business needs and user requirements

- **Communicate effectively**: Communicate SLA, SLI, and SLO targets and progress to all stakeholders clearly and concisely

Continuous improvement of SLAs, SLIs, and SLOs

Continuous improvement of SLAs, SLIs, and SLOs is a critical aspect of maintaining high-quality service delivery. As the needs and expectations of stakeholders change over time, it is essential to regularly review and adjust SLAs, SLIs, and SLOs to ensure that they remain relevant and effective. In this section, we will explore the importance of continuous improvement in maintaining SLAs, SLIs, and SLOs, as well as best practices for implementing and maintaining these improvements.

Some of the key topics we covered in this section are as follows:

- The importance of continuous improvement in SLAs, SLIs, and SLOs

- Reviewing and adjusting SLAs, SLIs, and SLOs over time

- Collecting and analyzing metrics to identify areas for improvement

- Implementing changes to improve SLAs, SLIs, and SLOs

- Automating processes for continuous improvement and monitoring

In this section, we explored the importance of building SLAs, SLIs, and SLOs with SRE principles, as well as the key principles and best practices for managing them. By implementing these principles and practices, you can ensure that your infrastructure is reliable, scalable, and efficient, while also meeting the needs of all stakeholders. Additionally, we saw how Terraform can be used to enforce SLAs, SLIs, and SLOs, making it an essential tool for managing your infrastructure. In the next section, we will explore how Terraform can be used to manage infrastructure at an enterprise scale.

How to run operations with Terraform

In this final section of this book, we will explore how to run operations with Terraform. As we have seen throughout this book, Terraform is a powerful tool for IaC and provides a way to define and manage infrastructure resources in a declarative manner. However, it is also important to understand how to use Terraform to manage and maintain infrastructure in production environments, and this section will cover best practices for doing so.

We will discuss the key considerations for running operations with Terraform, including managing state, version control, CI/CD, and using monitoring and alerting to maintain the health and performance of your infrastructure. By the end of this section, you will have a clear understanding of how to use Terraform to run operations in a scalable and reliable way.

Automating common operational tasks with Terraform

Automating common operational tasks with Terraform involves using Terraform to manage the infrastructure in production environments, automate recurring tasks, and ensure consistency across environments. This can include tasks such as deploying updates, scaling resources, and monitoring system health.

One key benefit of using Terraform for automation is the ability to apply changes quickly and reliably across the infrastructure. By defining IaC with Terraform, teams can ensure consistency and reliability in their infrastructure, reducing the likelihood of errors and increasing the speed of deployments.

Another benefit is the ability to monitor and maintain the infrastructure using Terraform. With the use of Terraform modules and providers, teams can automate tasks such as scaling, backups, and monitoring, reducing the workload on operations teams and increasing efficiency.

Overall, automating common operational tasks with Terraform can help teams streamline their operations, reduce downtime, and improve the reliability of their infrastructure. It also frees up resources for more strategic tasks and innovation.

Managing infrastructure changes with Terraform

As infrastructure grows and evolves, it's important to be able to manage changes effectively to ensure stability and minimize downtime. Terraform provides a powerful framework for managing infrastructure changes through its declarative language and state management.

One of the key benefits of Terraform is its ability to track changes to infrastructure over time. When infrastructure is deployed using Terraform, the state of the infrastructure is recorded in a file that can be used to manage and update the infrastructure over time. This allows you to easily track changes to the infrastructure and ensures that any changes are made in a controlled and repeatable way.

When making infrastructure changes, it's important to follow best practices to ensure that changes are made in a safe and controlled manner. One approach is to use a "plan, apply, and review" process. This involves creating a plan of the changes to be made, applying the changes, and then reviewing the results to ensure that they were applied correctly and did not introduce any unintended consequences.

Terraform also provides tools for managing changes across multiple environments, such as development, testing, and production. By using modules and workspaces, it's possible to manage changes consistently across different environments, while still allowing for environment-specific configurations.

Overall, Terraform provides a powerful framework for managing infrastructure changes and ensuring that changes are made in a safe and controlled way. By following best practices and leveraging the tools provided by Terraform, it's possible to manage infrastructure changes with confidence and minimize downtime and risk.

Monitoring and logging infrastructure with Terraform

Monitoring and logging infrastructure with Terraform is an essential part of any operational process. It helps to identify issues and take corrective actions before they escalate into critical problems. Terraform provides several tools and features that enable users to monitor and log their infrastructure.

One such tool is the Terraform provider for monitoring and logging services, which allows users to integrate their infrastructure monitoring and logging with their Terraform workflow. This provider supports various popular monitoring and logging services, including Datadog, Splunk, and CloudWatch.

By integrating monitoring and logging with Terraform, users can gain several benefits:

- They can automate the setup and configuration of monitoring and logging services for infrastructure
- They can track infrastructure changes over time and identify their impact on performance and availability
- They can get real-time alerts and notifications for critical events or incidents in the infrastructure
- They can correlate log data with metrics and traces to troubleshoot issues more effectively

To leverage the monitoring and logging provider, users can define the required resources in their Terraform configuration files, such as alerts, dashboards, and metrics. Terraform then takes care of creating and updating these resources in the respective monitoring and logging services.

Furthermore, Terraform allows users to create custom monitoring and logging solutions using open source tools and libraries. For example, users can use Terraform to deploy and configure Prometheus and Grafana for monitoring and visualization.

In summary, monitoring and logging infrastructure with Terraform is a critical part of any operational process. It enables users to automate the setup and configuration of monitoring and logging services, track infrastructure changes, and get real-time alerts and notifications for critical events.

Troubleshooting infrastructure issues with Terraform

As with any infrastructure or application, issues and incidents can arise in your AWS environment. When such incidents occur, it is important to quickly and efficiently troubleshoot and resolve the underlying issues. Terraform can be a valuable tool in this process, enabling you to identify and troubleshoot issues in your infrastructure configuration and apply fixes in a controlled, repeatable way:

- **Using Terraform state**: Terraform's state file provides a record of the current state of your infrastructure as it exists in the cloud. By examining the state file, you can identify differences between the desired and actual states of your infrastructure, which can help you pinpoint issues and take steps to address them.

- **Examining Terraform logs**: Terraform logs contain detailed information about the actions that Terraform is taking to manage your infrastructure. By examining these logs, you can gain insights into the specific steps that Terraform is taking, and identify any errors or issues that may be preventing your infrastructure from functioning as intended.

- **Using plan and apply commands**: Terraform's `plan` and `apply` commands allow you to preview and apply changes to your infrastructure configuration in a controlled manner. By using these commands, you can ensure that any changes you make to your infrastructure are applied in a safe and controlled manner, minimizing the risk of introducing new issues or errors.

- **Using Terraform modules**: Terraform modules can be used to simplify and standardize the troubleshooting and remediation process across different infrastructure components. By creating reusable modules for common infrastructure components, you can streamline the process of identifying and addressing issues and ensure that troubleshooting efforts are consistent and effective across your entire infrastructure.

- **Integrating with other tools and services**: Terraform can be integrated with other troubleshooting tools and services, such as AWS CloudWatch and AWS Systems Manager, to gain deeper insights into infrastructure issues and automate remediation processes. By leveraging these tools and services alongside Terraform, you can create a comprehensive infrastructure troubleshooting and remediation workflow that is both efficient and effective.

Scaling and managing infrastructure with Terraform

One of the primary benefits of using Terraform is its ability to manage and scale infrastructure in a consistent and repeatable manner. This includes scaling up or down resources based on changing demands, as well as managing the life cycle of infrastructure resources.

Some key considerations when scaling and managing infrastructure with Terraform are as follows:

- Using Terraform modules to standardize and simplify the management of infrastructure resources, such as EC2 instances, databases, and load balancers

- Leveraging Terraform's resource dependencies and life cycle management features to ensure that resources are provisioned and decommissioned in the correct order and that any associated data is preserved

- Designing infrastructure with scalability in mind, such as using auto-scaling groups and other techniques, to automatically add or remove resources based on changing demand

- Using Terraform's workspace feature to manage multiple environments, such as development, staging, and production, and to ensure that infrastructure changes are applied consistently across all environments

- Incorporating infrastructure monitoring and alerting into the scaling and management process to ensure that issues are detected and addressed promptly

- Using Terraform's version control features to track changes to infrastructure over time, and to roll back to previous configurations if necessary

- Regularly reviewing and updating infrastructure configurations to ensure that they remain optimized and aligned with business needs

By using Terraform to manage and scale infrastructure, organizations can ensure that their infrastructure is reliable, consistent, and easily managed, even as demands change and the business evolves.

In this section, we explored the various ways in which Terraform can be used to run operations on cloud infrastructure. We started by discussing how common operational tasks can be automated using Terraform, allowing for more efficient and streamlined management of infrastructure resources. Then, we examined how Terraform can be used to manage infrastructure changes, monitor and log infrastructure events, and troubleshoot infrastructure issues. Finally, we looked at how Terraform can be used to scale and manage infrastructure resources as needs change and evolve. With the help of Terraform, operations teams can gain greater control and visibility over their cloud infrastructure, leading to improved efficiency, security, and reliability.

Summary

In this final chapter, we explored how to achieve perfect infrastructure with Terraform in AWS. We started by discussing the key considerations for designing and developing infrastructure that meets stakeholder needs, achieves high availability and security, enables scalability, and maximizes efficiency. Then, we delved into the importance of continuous improvement and iteration, building SLAs/SLIs/SLOs with SRE principles, and how to run operations with Terraform.

We learned how to automate common operational tasks, manage infrastructure changes, monitor and log infrastructure, troubleshoot issues, and scale and manage infrastructure with Terraform. By leveraging Terraform's capabilities, we can simplify and standardize infrastructure management, achieve greater efficiency, and reduce the risk of human error.

With the knowledge and skills you've gained from this chapter, you will be well-equipped to build and manage perfect infrastructure in AWS with Terraform. From defining infrastructure requirements to establishing a design framework, implementing best practices, testing and validating infrastructure, and continuously improving infrastructure, this chapter provided a comprehensive guide to mastering Terraform in AWS.

Index

I

N

network security groups (NSGs) 156

O

Open Policy Agent (OPA) 164, 180
outputs 105

P

package managers
    Chocolatey on Windows 38
    Homebrew on OS X 37
    Linux 38
patterns and practices, IaC
    documentation 7
    modules and versions 6
    security and compliance 9
    source control and VCS 5
    testing 7-9
Payment Card Industry Data Security
    Standard (PCI DSS) 156, 188, 189
platform as a service (PaaS) 90
pre-compiled binary, manual installation
    Mac or Linux PATH configuration 36
    Windows PATH configuration 36
principle of least privilege (PoLP) 156
principles, IaC
    idempotency 4
    immutability 4, 5
projects
    defining 99
providers 21, 102, 103
provisioners 105, 106

R

Relational Database Service (RDS) 124, 157
resource names and tags
    used, for defining AWS projects
        or environments 100
resources 24, 102
return on investment (ROI) 69
role-based access control
    (RBAC) 151, 164, 186
root module 22

S

scalable infrastructure
    considerations, for designing 206
secure infrastructure, in Terraform
    building 192
    compliance checks, automating 193, 194
    least privilege, implementing
        with IAM policies 192
    secrets, storing securely 194
    secure network architectures, creating 193
    state, managing 195
security 195
    benefits, with Terraform 196
security and compliance 9
    compliance 10
    Identity and Access Management (IAM) 9
    secrets management 9
    security scanning 9
Semantic Versioning (SemVer) 155
Sentinel 180
serverless infrastructure 116
    designing and deploying, with
        Terraform 119-121

‹packt›

www.packtpub.com

Subscribe to our online digital library for full access to over 7,000 books and videos, as well as industry leading tools to help you plan your personal development and advance your career. For more information, please visit our website.

Why subscribe?

- Spend less time learning and more time coding with practical eBooks and Videos from over 4,000 industry professionals

- Improve your learning with Skill Plans built especially for you

- Get a free eBook or video every month

- Fully searchable for easy access to vital information

- Copy and paste, print, and bookmark content

Did you know that Packt offers eBook versions of every book published, with PDF and ePub files available? You can upgrade to the eBook version at packtpub.com and as a print book customer, you are entitled to a discount on the eBook copy. Get in touch with us at customercare@packtpub.com for more details.

At www.packtpub.com, you can also read a collection of free technical articles, sign up for a range of free newsletters, and receive exclusive discounts and offers on Packt books and eBooks.

Other Books You May Enjoy

If you enjoyed this book, you may be interested in these other books by Packt:

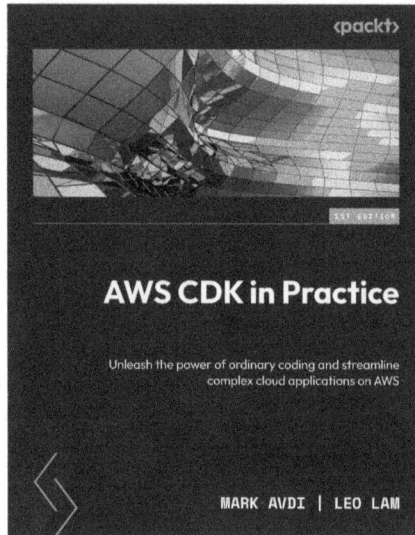

AWS CDK in Practice

Mark Avdi | Leo Lam

ISBN: 978-1-80181-239-9

- Turn containerized web applications into fully managed solutions
- Explore the benefits of building DevOps into everyday code with AWS CDK
- Uncover the potential of AWS services with CDK
- Create a serverless-focused local development environment
- Self-assemble projects with CI/CD and automated live testing
- Build the complete path from development to production with AWS CDK
- Become well versed in dealing with production issues through best practices

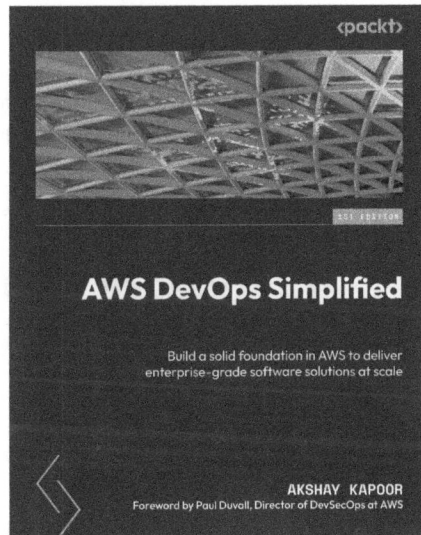

AWS DevOps Simplified

Akshay Kapoor

ISBN: 978-1-83763-446-0

- Develop a strong and practical understanding of AWS DevOps services
- Manage infrastructure on AWS using tools such as Packer and CDK
- Implement observability to bring key system behaviors to the surface
- Adopt the DevSecOps approach by integrating AWS and open source solutions
- Gain proficiency in using AWS container services for scalable software management
- Map your solution designs with AWS's Well-Architected Framework
- Discover how to manage multi-account, multi-Region AWS environments
- Learn how to organize your teams to boost collaboration

Packt is searching for authors like you

If you're interested in becoming an author for Packt, please visit authors.packtpub.com and apply today. We have worked with thousands of developers and tech professionals, just like you, to help them share their insight with the global tech community. You can make a general application, apply for a specific hot topic that we are recruiting an author for, or submit your own idea.

Share Your Thoughts

Now you've finished *Architecting AWS with Terraform*, we'd love to hear your thoughts! Scan the QR code below to go straight to the Amazon review page for this book and share your feedback or leave a review on the site that you purchased it from.

https://packt.link/r/1803248564

Your review is important to us and the tech community and will help us make sure we're delivering excellent quality content.

Download a free PDF copy of this book

Thanks for purchasing this book!

Do you like to read on the go but are unable to carry your print books everywhere?

Is your eBook purchase not compatible with the device of your choice?

Don't worry, now with every Packt book you get a DRM-free PDF version of that book at no cost.

Read anywhere, any place, on any device. Search, copy, and paste code from your favorite technical books directly into your application.

The perks don't stop there, you can get exclusive access to discounts, newsletters, and great free content in your inbox daily

Follow these simple steps to get the benefits:

1. Scan the QR code or visit the link below

https://packt.link/free-ebook/9781803248561

2. Submit your proof of purchase
3. That's it! We'll send your free PDF and other benefits to your email directly

www.ingramcontent.com/pod-product-compliance
Lightning Source LLC
Chambersburg PA
CBHW081059220326
41598CB00038B/7157